Response to
Occupational Health Hazards
A Historical Perspective

Response to Occupational Health Hazards

A Historical Perspective

Jacqueline Karnell Corn

VNR VAN NOSTRAND REINHOLD
_____ New York

Library of Congress Catalog Card Number 91-47173
ISBN 0-442-00488-5

Manufactured in the United States of America

Published by Van Nostrand Reinhold
115 Fifth Avenue
New York, New York 10003

Chapman and Hall
2-6 Boundary Row
London, SE1 8HN, England

Thomas Nelson Australia
102 Dodds Street
South Melbourne 3205
Victoria, Australia

Nelson Canada
1120 Birchmount Road
Scarborough, Ontario M1K 5G4, Canada

16   15   14   13   12   11   10 9   8   7   6   5   4   3   2   1

Library of Congress Cataloging-in-Publication Data
Corn, Jacqueline K.
    Response to occupational health hazards: a historical perspective/
Jacqueline Karnell Corn.
        p.   cm.
    Includes bibliographical references and index.
    ISBN 0-442-00488-5
    1. Industrial hygiene—History.   I. Title.
RC967.C62   1992
363.11'0973—dc20   91-47173
                                                                        CIP

Mankind does not of a sudden become sinless, but he knows little of history who thinks that progress is downhill, not up. We do swing down sometimes but not as far as we swung the last time, and when we swing up again it is to go higher than we went before.

Alice Hamilton

This book is dedicated to the many professionals in occupational health who continue to strive for improved working conditions and reduced injury and illness on the job.

Contents

Preface

Once it was hoped that scientific approaches to occupational health (for example, definition and understanding of disease) and the engineering solutions of setting and adhering to safe factory standards would provide humane solutions to the problem of industrial diseases. We now understand that solutions are more complicated. Hazards also pose political and social challenges, and occupational health is affected by a wide range of social factors.

Today, uncertainties and complexities involved in identifying and assessing work-related diseases, rapid proliferation of new hazards, better ability to quantitate, the increasing role of regulation, and interest generated by new laws, the press, labor, and public interest groups have all placed the increasingly volatile issues associated with occupational health in the public arena. Current controversy about occupational health focuses on the selection of priorities, policy decisions, and regulatory activities for those hazards to be controlled, and the associated degree of control.

Occupational illness is a social problem. Yet only a small part of the historical literature on industrialization, the labor movement, or even the history of medicine is concerned with occupational health. Although the exact number of deaths and injuries related to work is disputed, the Office of Technology Assessment has estimated about 6,000 deaths annually due to injuries and, depending on how many injuries were counted, between 2.5 million and 11.3 million nonfatal occupational injuries.[1] Even less reliable quantitative information on occupational illness exists. Bureau of Labor Statistics annual surveys, based on employer records, underestimate the magnitude of the occupational disease problem. The most commonly quoted numbers include 100,000 deaths due to illness and 390,000 cases of illness each year resulting from working conditions.[2]

What are the effects of industry on the health of employees? The first and most obvious effect is that of accidental death or injury, although the risk of accidents at work differs from one industry to another. Frequency rates (disabling injuries per million hours worked) vary, as do severity rates (time charges per million hours worked). In addition to traumatic injuries, workers face the risk of developing occupational diseases. Reliable statistics still do not exist for the occurrence of occupational diseases.

Many occupational diseases cannot be distinguished from diseases that result from other than occupational exposures. Nonspecific upper respiratory disease, bronchitis, and lung cancer can all be related to smoking, air pollution, or exposure to certain contaminants in the work environment.

There are also manifestations of occupational disease specific to certain chemical agents, for example, silicosis, a fibrosis of the lung caused by dust containing free silica, and mesothelioma, a form of malignancy uniquely related to asbestos dust in the lung. The long period of incubation required to develop many occupationally related diseases is another reason for uncertainty in the statistics of occupational disease. Years of exposure to an agent often pass before disease symptoms develop. Lung cancer related to exposure to coke oven emissions or asbestos can require up to 20 years to develop. A worker may change jobs during this long incubation period, making it unlikely that a physician will relate the disease to earlier exposure to the disease agent at work.

These reasons for inaccurate occupational disease statistics leave us with estimates that range from 25,000 to 100,000 new industrial disease cases yearly, including cases of dermatosis, perhaps the most prevalent industrial disease.

The psychosocial stresses in the work environment have not been emphasized in the United States. One need only read Studs Terkel's *Working* to realize how widespread boredom with work is in the United States and how little satisfaction people derive from their jobs.[3] Concern for the mental stresses of work will surely increase in the near future.

Occupational illness is not new, but the magnitude and nature of the problems and approaches to solving them have changed over the years. Technology changes, and so do social values and norms. This book utilizes historical methods to analyze and describe how the changes in approach to occupational health issues are related to social values, perception of risk, definition of disease, and scientific and technological developments. As these factors alter, attitudes and actions related to occupational health hazards also change.

Occupational health can be traced back to early observations of the relation between work and disease. In antiquity, work in mining, metallurgy, pottery making, and glass making and the utilization of dyes and ceramic glazes often caused sickness. These early observations of the relation between illness and occupation were severely limited by lack of knowledge, the small number of workers at risk, and social acceptance of work-induced illness.[4] The benefits of work, when contrasted with the risks, seemed to outweigh the known hazards. It took centuries to replace the belief that disease was an acceptable and unavoidable by-product of work with the idea that prevention and control of hazards on the job could minimize or even eliminate risk. Indeed, dangerous and unhealthy conditions of work were accepted, until quite recently, as necessary outgrowths of industrial expansion.

This book examines the historical nature of occupational health issues by focusing on modern United States approaches to the definition, recognition, and control of occupational disease. It is about how American society perceived occupational health problems and proceeded to solve them or in some cases to ignore them. The book raises questions about perception of risk and its relation to political and social values. In the twentieth century new attitudes and actions toward the health of working men and women appeared. They were related to technology and science, that is, improved methods of detection and control, better diagnostic techniques, improved epidemiological methodology, better understanding of disease and methods of preventing disease, as well as to social attitudes, that is, social response to disease, changed perception of disease, and assumption of responsibility by government, labor, and management for the health of workers.

Case studies are utilized to place current occupational health issues in social and scientific context and to illustrate the complex situation that occurs when industrial development affects human health. Because occupational health is affected by such a wide range of social and technological factors, the case studies in this book are designed to illustrate many of the social, scientific, and political factors that led to recognition of industrial diseases and mobilization to intervene to control or eliminate them.

Chapter 1 presents a short historical overview of the growth of federal responsibility in the United States for occupational health. Chapter 2 provides historical perspective concerning risk assessment and occupational health policy, and chapter 3 is an overview of occupational health standards. Chapters 4 through 8 are case studies of lead, asbestos, silicosis, vinyl chloride, and byssinosis, respectively, designed to illustrate a variety of themes in occupational health history.

Chapter 4 originally appeared in the *Milbank Quarterly: Health and Society,* Winter 1975. Chapters 5 and 8 appeared in the *American Journal of Industrial Medicine,* volume 2, 1981 and volume 11, 1987 respectively. The original article on asbestos that chapter 8 is based on was co-authored with Jennifer Starr. Chapter 6 appeared in *The Journal of The American Industrial Hygiene Association* in 1980. Chapter 7 appeared in *The Journal of Public Health Policy* in 1984. Each of the articles has been brought up to date for publication in this book. I wish to acknowledge and thank these journals for permission to reprint the articles.

I have drawn upon history because it presents a perspective necessary to clarify issues, as well as insights that can improve the way we choose options, set priorities, and make occupational health policy decisions. Understanding how we defined and sought to control occupational hazards in the past and placing them in historical and social context can clarify current issues.

References

1. *Preventing Illness and Injury in the Workplace*. OTA-H-256. Washington; U.S. Congress Office of Technology Assessment, 1985.
2. Ibid., 37.
3. Terkel, Studs. *Working*. New York: Pantheon Books, 1974.
4. Rosen, George. *The History of Miner's Diseases*. New York: Schuman's, 1943.

Response to Occupational Health Hazards
A Historical Perspective

Chapter 1

Historical Perspective on Government Responsibility for Occupational Health

Concern for occupational health coupled with effective action is relatively recent. Looking backward, it is clear that three changes occurred during the twentieth century that help to explain the history of occupational health in the United States. First, the idea that prevention and control of hazards on the job can minimize and even eliminate risk slowly replaced the opinion that accidents and diseases are unavoidable by-products of work. Second, the assumption by government for the health and safety of workers replaced the nineteenth-century idea of laissez-faire, which had allowed for lack of significant social conscience pertaining to dangerous conditions of work and minimal activity to protect men and women at work. The third and most recent change involves the potent idea that working men and women have the right to know about hazards on the job and to act to better their own working conditions, including those related to job health and safety.

The continuous interplay between science and politics and the gap between awareness and action to eliminate hazards are two recurring themes in occupational health history. In spite of the interesting subject matter, only a small part of American historical literature on industrialization, the labor movement, medicine, or even public health is concerned with occupational health.[1, 1A–1G] A few historians have recently published books and articles in this area, for example: David Rosner and Gerald Markowitz edited *Dying for Work* and wrote *Deadly Dust;* Alan Derickson, *Workers' Health, Workers' Democracy;* and Jacqueline Corn, *Protecting the Health of Workers* and *Environment and Health in Nineteenth Century America*. Rosner, Markowitz, Corn, Derickson, Fry, and Sellers have published in journals on the history of occupational health.

1

Occupational health decisions, similar to public health decisions, have two elements. The element of value determines the degree or extent of health a society wants and is willing to pay for. The scientific element determines how much knowledge a society has upon which to base its occupational health policy. Occupational health professionals once naively acted on the belief that science and scientific approaches alone would provide solutions to the problems associated with occupational diseases that could lead to safe workplaces.[2] Today it is clear to all that industrial hazards also pose political, social, and economic problems, complicated by a perception of hazards that continually alters in response to changing scientific understanding interwoven with social, political, and economic conditions.

The significant political, social, economic, and technological changes of the twentieth century help to explain developments in occupational health. Three critical periods in American history—1900–1917, 1935–1946, and 1968–1970—illustrate these changes.

1900–1917

In the late nineteenth and early twentieth centuries, an economic revolution dominated and transformed American life. Between the Civil War and World War I, the United States developed into a leading industrial and manufacturing nation. A number of factors account for this. They include:

1. The application of technology to manufacturing and extraction of raw materials
2. The discovery and utilization of natural resources (iron, coal, natural gas, copper, etc.)
3. A large supply of labor
4. Construction of a transportation system
5. Growth of foreign and domestic markets
6. Creation of capital
7. Favorable governmental policies such as protective tariffs and indirect subsidies

Although all of these factors helped to create an industrial giant, the first factor, the application of technology, dominated the others. Dynamic technical changes, the introduction of new processes, and the ever-increasing new sources and new uses of mechanical power affected every aspect of American life. Figure 1-1 illustrates the rapidity of industrial growth with the index of manufacturing production from 1860 to 1915.

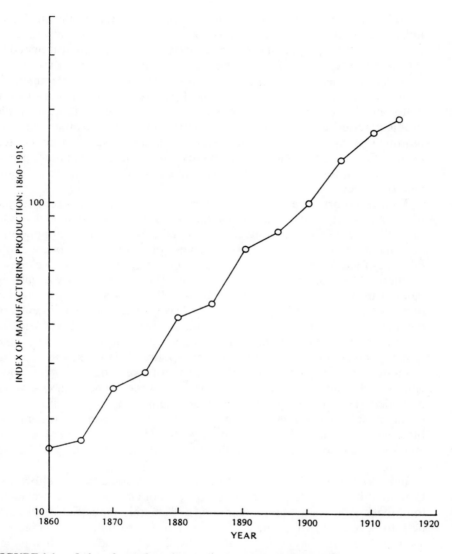

FIGURE 1-1. Index of manufacturing production: 1860–1915. Source: *Historical Statistic of the U.S.: Colonial Times to 1970*, Part 2, Series P13-28 (Washington: U.S. Departments of Commerce and Interior,), 667.

Industrialization and urbanization profoundly affected the health of individuals in factories, mines, and cities. Each new process, machine, or technological development created hazards. At the time few responded to the health effects that industrialization had on individuals exposed to unsafe and unhealthy working conditions. The rising toll of deaths from industrial accidents and diseases caused by exposure to toxic industrial materials merited attention but was only slowly recognized as a solvable problem. Often ignored working conditions in factories and mines led to an inordinate number of occupational accidents and diseases. Workers, often aware of the hazards to which they were exposed, faced even more pressing problems, for example, low wages, long and arduous hours of work, and job insecurity.

By the Progressive Era, the problems associated with occupational health and safety had reached epidemic proportions that demanded attention. The expansion of mining and manufacturing, growth of monopolies, consolidation of industries, and the genesis of industrial giants created a new set of social problems, among them an unsafe, unhealthy work environment. The inordinate number of occupational exposures was largely ignored in part because meager social legislation existed in an America dominated by the philosophy of social Darwinism and a business ethic of competitiveness. Hazards to health and safety were many, primitive in nature, and large in magnitude. The long-known, poorly defined occupational diseases of silicosis, plumbism, phosphorus poisoning, and mercurialism occurred frequently in workplaces that lacked adequate ventilation and sanitary facilities and where long and burdensome hours of work prevailed. The necessary description and definition of these hazards by means of scientific studies and surveys and the education of workers and the general public had not yet taken place. The attitude prevailed that employers had no more responsibility toward workers than they wished to assume.

Industrial accidents caused by the lack of an organized safety effort, the general attitude of irresponsibility on the part of employers, the prevalent long work day, unorganized workers, and chaotic industrial conditions help to account for high incidence of death due to industrial accidents. Table 1-1 reveals that approximately 23,000 known industrial deaths occurred in a total workforce of 38,000,000 during one year, 1913.[3]

Government and private organizations responded slowly to the health problems introduced or aggravated by the growth of industry. If statutes and laws are viewed as instruments of public policy, then broad public policy in occupational safety and health did not crystallize until many years later.

The growth of mass-circulation newspapers and national magazines

TABLE 1-1.   Estimates of Fatal Industrial Accidents in the United States, 1913

Industry Group (Males)	Number of Employees	Fatal Indus. Accidents	Rate per 1,000
Metal mining	170,000	680	4.00
Coal mining	750,000	2,625	3.50
Fisheries	150,000	450	3.00
Navigation	150,000	450	3.00
Railroad employees	1,750,000	4,200	2.40
Electricians (light and power)	68,000	153	2.25
Navy and marine corps	62,000	115	1.85
Quarrying	150,000	255	1.70
Lumber industry	531,000	797	1.50
Soldiers, U.S. Army	73,000	109	1.49
Building and construction	1,500,000	1,875	1.25
Draymen, teamsters, etc.	686,000	686	1.00
Street railway employees	320,000	320	1.00
Watchmen, policemen, firemen	200,000	150	0.75
Telephone and telegraph (including linemen)	245,000	123	0.50
Agricultural pursuits, including forestry and animal husbandry	12,000,000	4,200	0.35
Manufacturing (general)	7,277,000	1,819	0.25
All other occupied males	4,678,000	3,508	0.75
All occupied males	30,760,000	22,515	0.73
All occupied females	7,200,000	540	0.075

SOURCE: *Industrial Accident Statistics*, Bulletin 157 (Washington: United States Department of Labor, Bureau of Labor Statistics, 1915), 5.

helped to forge a national interest in workers' safety and health and to alert many Americans to the issues of health and safety associated with industry. In 1907 a mine disaster in Monongah, West Virginia, killed 362 coal miners. This widely publicized disaster shocked the American public and led to the creation of the United States Bureau of Mines in 1910. One purpose of the Bureau of Mines was to promote mine safety.[4]

Impetus for change during the Progressive Era came from the humanitarian segment of the progressive movement. In response to the challenge of unacceptable injuries, pioneers in the field of occupational health and safety initiated a safety movement to control industrial accidents and an industrial hygiene movement to control occupational disease. They also sought workmen's compensation laws to ameliorate the plight of injured workers and their dependents by assuring at least partial compensation for lost wages. In 1911 Wisconsin became the first state to establish a workmen's compensation program. Most other states followed the Wisconsin example. Wisconsin also developed the first permanent state industrial commission with authority to establish and enforce safety and health regulations. Wisconsin based its regulations on comments from both labor and management.[5]

The Pittsburgh Survey, sponsored by the Russell Sage Foundation, depicted living conditions and working conditions in Allegheny County, Pennsylvania. Crystal Eastman undertook the first systematic investigation of accidents occurring in a representative period (1906–1907), in a representative district (Allegheny County, Pennsylvania) for the Pittsburgh Survey. To state the problem and illustrate her point, she presented four case studies of injured workers. Eastman's purpose was "to determine in cases studied, (1) What are the indications as to responsibility? (2) What material loss and privation, if any, resulted to the injured workmen and their families?"[6] She considered responsibility for work-related accidents and its bearing on the just determination of economic loss. From July 1, 1906 to June 30, 1907, work accidents killed 326 men in Allegheny County, Pennsylvania. Eighty-four percent of the men killed were 45 years old or under. Fifty-eight percent of the men killed were not yet over 30.[7] Figure 1-2, the "Death Calendar" Eastman included in her book, vividly illustrates the point she wished to make. Deaths were marked by red Xs.

Eastman sought enactment of workmen's compensation laws because responsibility for accidents was due, as her study proved, to circumstances beyond the control of the workers, and survivors usually received little or no compensation.

The prevalent attitude toward injured workers was that employers had no more responsibility for employees than they wished to assume. If an employee or his survivor sought compensation in the court, common law defenses stacked the odds in the favor of the employer. These common law defenses included

1. Fellow servant: if an injury resulted from another worker's negligence, the injured employee could not collect compensation from the employer.
2. Assumption of risk: the employee assumes the risk inherent in a job that he knew of or should have known about.
3. Contributory negligence: if the employee was negligent in any way, he could not collect anything, regardless of the employer's negligence.

Alice Hamilton, a physician, vividly described the deplorable state of occupational health in the United States during this same era.[8] Dr. Hamilton described herself as a pioneer in the unexplored field of industrial medicine. During her career she worked to change, for the better, the

DEATH CALENDAR IN INDUSTRY
FOR ALLEGHENY COUNTY

JULY 1906

S	M	T	W	T	F	S
1	2	3	4	5	6	7
8	9	10	11	12	13	14
15	16	17	18	19	20	21
22	23	24	25	26	27	28
29	30	31				

35

AUGUST 1906

S	M	T	W	T	F	S
			1	2	3	4
5	6	7	8	9	10	11
12	13	14	15	16	17	18
19	20	21	22	23	24	25
26	27	28	29	30	31	

45

SEPTEMBER 1906

S	M	T	W	T	F	S
						1
2	3	4	5	6	7	8
9	10	11	12	13	14	15
16	17	18	19	20	21	22
23	24	25	26	27	28	29
30						

37

OCTOBER 1906

S	M	T	W	T	F	S
	1	2	3	4	5	6
7	8	9	10	11	12	13
14	15	16	17	18	19	20
21	22	23	24	25	26	27
28	29	30	31			

35

NOVEMBER 1906

S	M	T	W	T	F	S
				1	2	3
4	5	6	7	8	9	10
11	12	13	14	15	16	17
18	19	20	21	22	23	24
25	26	27	28	29	30	

54

DECEMBER 1906

S	M	T	W	T	F	S
						1
2	3	4	5	6	7	8
9	10	11	12	13	14	15
16	17	18	19	20	21	22
23	24	25	26	27	28	29
30	31					

48

JANUARY 1907

S	M	T	W	T	F	S
		1	2	3	4	5
6	7	8	9	10	11	12
13	14	15	16	17	18	19
20	21	22	23	24	25	26
27	28	29	30	31		

60

FEBRUARY 1907

S	M	T	W	T	F	S
					1	2
3	4	5	6	7	8	9
10	11	12	13	14	15	16
17	18	19	20	21	22	23
24	25	26	27	28		

36

MARCH 1907

S	M	T	W	T	F	S
					1	2
3	4	5	6	7	8	9
10	11	12	13	14	15	16
17	18	19	20	21	22	23
24	25	26	27	28	29	30
31						

43

APRIL 1907

S	M	T	W	T	F	S
	1	2	3	4	5	6
7	8	9	10	11	12	13
14	15	16	17	18	19	20
21	22	23	24	25	26	27
28	29	30				

51

MAY 1907

S	M	T	W	T	F	S
			1	2	3	4
5	6	7	8	9	10	11
12	13	14	15	16	17	18
19	20	21	22	23	24	25
26	27	28	29	30	31	

42

JUNE 1907

S	M	T	W	T	F	S
						1
2	3	4	5	6	7	8
9	10	11	12	13	14	15
16	17	18	19	20	21	22
23	24	25	26	27	28	29
30						

42

FIGURE 1-2. Death calendar. The Pittsburgh Survey (started in 1906) was the first industrial accident study. The results were shocking: 326 men killed in in-plant accidents in a 12-month period, an average of 10 men killed every week in the year.

health status of American workers. Alice Hamilton's most valuable contributions included descriptions of the extent of industrial disease in the United States. She said in her autobiography that in the early twentieth century most of the literature on the subject of industrial disease was European.

> In those countries industrial medicine was a recognized branch of medical sciences: in my own country it did not exist. When I talked to medical friends about the strange silence on the subject in American medical magazines and textbooks, I gained the impression that here was a subject tainted with socialism or with feminine sentimentality for the poor. The American Medical Association had never had a meeting dedicated to this subject, and except for a few surgeons attached to large companies operating steel mills, or railways, or coal mines, there were no medical men in the State of Illinois who specialized in the field of industrial medicine.[9]

In 1910 Governor Deneen of Illinois appointed an occupational disease commission to inquire into the extent of industrial disease in his state. It was the first such survey undertaken in the United States. The commission included Alice Hamilton, whose investigations centered on lead hazards. In 1912 she made a similar survey for the United States government. During her travels around the United States, she visited lead smelters, storage battery plants, and other hazardous workplaces. Alice Hamilton's only authority to enter plants was persuasion. In spite of her lack of authority and the minimal attention to occupational diseases during this period, Hamilton made valuable contributions to the field of occupational health. Her surveys helped to define and describe occupational hazards, including the extent of occupational disease during the Progressive Era.

The federal government entered the field of industrial health early in the twentieth century, albeit with minimal efforts. Its Bureau of Labor began to publish studies of the dusty trades in 1903. A 1910 study published by the Bureau of Labor and written by John B. Andrews on the horrors of "phossy jaw" (phosphorus necrosis), a painful, disfiguring, and often fatal disease of workers in the white phosphorus match industry, led to the Esch Act of 1912, which placed a prohibitive tax on white phosphorus. At the same time the Diamond Match Company released its patent on a substitute for white phosphorus.[10]

When Congress created the Department of Labor in 1913 to "improve working conditions," a duty of newly appointed Secretary of Labor William B. Wilson was to report on industrial diseases and accidents. The Bureau of Labor Statistics within the Department of Labor began to compile accident statistics and published *Industrial Accident Statistics*.[11]

Its author, Frederick L. Hoffman, noted the lack of trustworthy industrial accident statistics in the United States, in part because of the absence of uniform requirements for reporting accidents in the various states.

In 1914 federal responsibility for industrial safety and health was also placed in the Office of Industrial Hygiene and Sanitation in the United States Public Health Service. In its early years, the Office of Industrial Hygiene and Sanitation engaged in laboratory and field research. Table 1-2 is a list of reports resulting from both Public Health Service and Department of Labor field research activities.

Congress created two other agencies concerned with worker health and safety: the United States Bureau of Mines (noted earlier in this chapter) and the United States Children's Bureau.

In reality the federal agencies did little to protect the health and safety of working men and women. They collected statistics, pursued research and field studies, and endeavored to disseminate their findings. The prevailing philosophy remained that protection of industrial workers was the responsibility of state and local public health agencies. Protection meant, to most state and local health agencies, matters of sanitation, employment of women and children, and compensation to employees following an accident. Wisconsin, an exception to this rule, created the first permanent industrial commission to develop and enforce safety and health regulations.

Other state governments passed minimal legislation affecting worker health. Some of the legislation was declared unconstitutional; for example, Maryland legislation passed in 1902 for benefits to workmen injured in the course of employment was declared unconstitutional in 1904. Illinois enacted an eight-hour day for children under age 16. Washington, followed by California, Nevada, Illinois, Ohio, Wisconsin, Kansas, Massachusetts, New Hampshire, and New York, passed a compulsory workmen's compensation law in 1911. In 1913 New York and Ohio established state industrial hygiene units staffed by physicians and engineers. However, occupational health measures to safeguard the health of working men and women remained minimal.

In the private sector, in 1906 the American Association of Labor Legislation organized to conduct investigations, hold national conferences, publish reports, draft bills, and secure enactment into law of progressive standards. In 1910 it held its first National Conference on Industrial Diseases. In 1913 the National Council for Industrial Safety (later called the National Safety Council) came into existence. The Industrial Hygiene Section of the American Public Health Association, organized in 1914, was among the few national, private organizations directly interested in occupational health. The following year the Conference Board of Physicians in

TABLE 1-2.   Reports from Industrial Accidents and Hygiene Series

Report	Date
Deaths from Lead Poisoning, 1925–27	June 1929
Health Survey of the Printing Trades, 1922–25	March 1927
Phosphorus Necrosis in Manufacture of Fireworks and Preparation of Phosphorus	May 1926
Survey of Hygienic Conditions in the Printing Trades	September 1925
The Problem of Dust Phthisis in the Granite-Stone Industry	May 1922
Industrial Accidents (Poisonings) in Making Coal Tar Dyes and Intermediates	April 1921
Carbon Monoxide Poisoning	December 1921
Causes of Death by Occupation	February 1930
Preventable Death in Cotton Manufacturing Industry	October 1919
Lead Poisoning in the Smelting and Refining of Lead	February 1919
Women in the Lead Industries	February 1919
Lead Poisons in the Manufacture of Storage Batteries	December 1915
Industrial Poisonings Used in the Rubber Industry	October 1915
Mortality from Respiratory Diseases in Dusty Trades (Inorganic Dusts)	June 1918

Industry was established to advise the National Industrial Conference Board, and in 1916 the American Association of Industrial Physicians and Surgeons held its first annual meeting in Detroit.

The same period witnessed publication of the first modern American texts concerning occupational health. For example, *The Modern Factory* by George H. Price, *Diseases of Occupation and Vocational Hygiene* edited by George M. Kober and William C. Hanson, and *Industrial Medicine and Surgery* by Harry E. Mock.

Harvard offered instruction in industrial hygiene, and the first occupational disease clinic was established at the Cornell Medical Center Outpatient Department in New York City. The first union-sponsored type of medical care plan was started by the International Ladies Garment Workers Union in 1913, and in 1917 the Union Health Center was incorporated.

The activities of the Progressive Era can be viewed as an indication of change, marking the beginning of a long, difficult, and significant road toward recognition of dangerous and unhealthy working conditions. The response of both the public and private sectors included early but minimal governmental assumption for worker health and safety. *Minimal* cannot be stressed too much.

A literary comment on this era appears in Sholem Asch's *East River*. He wrote about the Triangle fire that occurred in the New York garment industry on March 25, 1911.

More than one hundred and fifty girls lost their lives in the fire. They were buried at mass funerals; the Jewish girls in the Jewish cemeteries; the Christian girls in Christian cemeteries. The survivors soon began to search for work in other factories. The wave of excitement and anger that swept through the city and all through the country didn't last very long. A commission was appointed to investigate fire hazards in the state's garment factories. Some bills were introduced into the Assembly. There were heated debates; some measures were adopted, others defeated. When it was all over, everything in the needle industries remained the same.[12]

The advent of World War I precipitated a crisis in industrial health and safety and stimulated some activity in war industries. Congress initiated the Working Conditions Service to inspect war production sites, advise companies on reducing hazards, and help states develop and enforce safety and health standards. When the war ended, the Working Conditions Service also ended.

The period following World War I was one of extreme conservatism. The Harding-Coolidge-Hoover era assisted and encouraged business enterprise by its policies of high tariffs, laissez-faire, suspension of the small amount of regulatory legislation that existed, reduction of taxation, and repression of labor. Occupational health activity was not, as one might guess, commensurate with the expansion of commerce and industry in the 1920s. Very little progress occurred toward making workplaces safe and healthy until the 1930s, when New Deal legislation stimulated the field of occupational hygiene.[13]

1935–1946: THE NEW DEAL

In October 1929 the stock market crashed. By 1932, when a quarter of the labor force was unemployed, neither the federal, state, nor local governments could provide adequate help. Franklin D. Roosevelt took the oath of office as the Great Depression reached its lowest level.

The New Deal was a program designed to rescue the United States from the worst depression in its history. A guiding principle of New Deal legislation enacted in the 1930s was government responsibility for human welfare. The Roosevelt administration set forth a program of reform and recovery. Its keynote was a new deal for the forgotten man. The New Deal achieved a major change in the relationship of government to society. Within the first year of Roosevelt's presidency, Congress enacted far-reaching social, labor, and economic legislation. Some of the laws directly affected working conditions, including safety and health. During this period the genesis of the growing federal assumption of responsibility for the health and safety of working men and women appeared.

Advances in social legislation during the New Deal spilled into the field of occupational health and led to renewed interest in worker safety and health, although responsibility for safeguarding the health of workers remained chiefly in state and local governments. The federal system allowed considerable freedom for each state to pursue its own industrial health and safety policies. The federal agencies concerned with occupational health engaged primarily in collecting and disseminating of information, conducting field studies and laboratory research, and protecting the health and safety of federal employees. As noted earlier, the states dealt chiefly with safety, sanitation, employment of women and children, and compensation of employees following accidents.

New federal legislation affecting workers during this period began to increase the role of the federal government in protecting people on the job. The National Industrial Recovery Act (declared unconstitutional in 1935) set up industrial codes for fair competition, including the regulation of hours and of safe and healthful working conditions. Four other laws augmented the role of the federal government to protect workers.

1. The Social Security Act (1935) allocated funds to the Public Health Service for research and grants-in-aid to states for public health work, including industrial hygiene.
2. The Public Contracts Act (Walsh-Healey, 1936) established labor standards on government contracts, including requirements for the safety and health of workers.
3. The Fair Labor Standards Act (1938) set a minimum wage, banned exploitive child labor, and gave the United States Department of Labor power to ban workers under the age of 18 from dangerous occupations.
4. The National Labor Relations Act (1935) guaranteed workers the right to organize and bargain collectively.

When Congress passed the National Labor Relations Act (the Wagner Act) guaranteeing workers the right to organize and bargain collectively and establishing a permanent board to supervise elections for union representatives (the National Labor Relations Board or NLRB), it allowed the Congress of Industrial Organizations (CIO) to actively organize workers. Organizing drives led to growth in trade union membership, which impacted health and safety by allowing organized labor to pursue aims of health and safety on the job. Labor struggles, more often than not, centered on wages, hours, and working conditions, health and safety included.

In 1934 Secretary of Labor Frances Perkins established the Division of Labor Standards to promote worker safety and health, "the first link in the continuous historical chain that led to the establishment of the Occupational Safety and Health Administration (OSHA)."[14]

The Division of Labor Standards enlarged the Department of Labor's role in promoting job safety and health. Its mandate, to improve working conditions, emphasized health and safety. In keeping with the times, the Division of Labor Standards sought to expand the federal role of improving labor standards through assistance to the states. Most workers were subject to state laws. The Division of Labor Standards worked with states to promote the idea of uniform labor laws, helped state legislatures draft new laws, organized conferences on labor legislation, helped prepare factory inspection manuals, and improved factory inspection by planning training courses for the inspectors. It helped unions set up safety and health programs and encouraged safety education. It also worked with safety organizations and industrial associations.

Although the idea still prevailed that progress in industrial hygiene could be made only through efforts on the part of state and local governments, few states had addressed industrial hygiene problems until the Social Security Act allocated federal funds to them for occupational health purposes. The Social Security funds were earmarked for the states by the United States Public Health Service. In an effort to channel these funds, the Public Health Service inaugurated a program to develop industrial hygiene in state health departments.[15] The money caused a burst of activity and led to the establishment of a number of state industrial hygiene units.

Prior to 1935 only five state health departments (Connecticut, Maryland, Mississippi, Ohio, and Rhode Island) conducted industrial hygiene activities, and they were of a limited nature. Massachusetts and New York had industrial hygiene units within their state labor departments. By 1936 17 industrial hygiene units existed in state and local health departments. By 1938 the number rose to 26. The 1936 budget of $100,000 for industrial hygiene climbed to nearly $750,000 in 1938.[16]

Table 1-3 indicates the states with industrial hygiene units in 1938. It also shows the number of workers; number of workers in mining, manufacturing, and mechanical industries; number of industrial hygiene personnel; and the amount of money budgeted for industrial hygiene.

In 1938 the Division of Industrial Hygiene in the Public Health Service, headed by Royd R. Sayers, had 59 industrial hygiene specialists, including medical and engineering officers. Three of these officers devoted their entire working time to assisting industrial hygiene units in the states.

The Division of Industrial Hygiene had taken on a monumental task, to implement the industrial hygiene provisions of the Social Security Act. The division needed to address both administrative and technical problems. Perhaps the most difficult administrative problem was lack of trained

TABLE 1-3. States with Industrial Hygiene Units, 1938

State	Number of Workers	Number of Workers in Manufacturing, Mechanical, and Mining Industries	Industrial Hygiene Personnel				Annual Budget
			Medical	Engineering	Technical	Clerical	
Alabama	1,026,320	220,378	1	0	0	0	5,600
California	2,500,969	673,646	1	1	1	1	22,020
Connecticut	677,292	337,445	1	1	3	2	25,557
Illinois	3,184,876	1,164,979	2	2	3	2	35,400
Indiana	1,251,127	464,549	1	1	0	1	11,500
Iowa	912,823	167,147	0	2	0	1	8,350
Kansas	694,276	132,662	0	1	1	2	10,820
Maryland	672,902	228,599	1	2	3	2	12,900
Massachusetts	1,814,422	840,300	0	1	2	2	19,200
Michigan	1,927,498	860,164	1	6	5	4	45,500
Mississippi	844,887	79,318	0	0	1	1	9,100
Missouri	1,458,054	390,399	0	3	0	1	13,005
New Hampshire	192,671	89,461	0	0	0	1	8,380
New York	5,523,085	1,995,924	7	7	12	7	118,614
North Carolina	1,141,129	289,917	2	2	1	1	27,500
Ohio	2,615,938	1,094,650	3	1	2	2	25,300
Pennsylvania	3,722,428	1,796,944	2	9	0	1	56,490
Rhode Island	293,168	164,304	2	3	1	1	23,000
South Carolina	687,721	146,344	1	1	0	1	10,300
Tennessee	948,209	213,077	1	1	0	0	11,000
Texas	2,237,118	395,802	1	3	1	1	24,290
Vermont	141,190	42,851	1	1	0	1	11,000
Virginia	880,276	221,539	1	1	1	1	19,000
Washington	664,813	181,765	0	1	0	1	6,200
West Virginia	570,459	242,115	1	0	0	1	17,500
Wisconsin	1,129,546	380,229	1	2	0	1	11,000
	37,723,197	12,814,508	32	54	37	38	589,026

SOURCE: J. J. Bloomfield, "Development of Industrial Hygiene in the United States," *American Journal of Public Health* 28: (December 1938) 1391.

14

personnel. The Division of Industrial Hygiene began in-service training courses for people entering the field of occupational health in the various states. A number of states also used Social Security funds to train people at university training centers at Harvard, Michigan, Johns Hopkins, Columbia, California, and Yale. The Public Health Service conducted training seminars for physicians, chemists, and engineers in 1936 and 1937.

Administrative problems confronting the new state industrial hygiene units included inadequate funding, lack of interest in occupational health and safety by state legislators, little or no accurate reporting of occupational disease, few surveys of industry to define existing hazards, and lack of what at the time were called "good practice codes" for the elimination of hazardous conditions at work. Furthermore, the idea persisted that occupational health professionals should only show industry how to solve its own problems. Setting and enforcing standards had not yet been accepted as the philosophical basis for creating healthful and safe working conditions. The task of the occupational health professionals was to "persuade" industry to solve its own health problems, aid industry to evaluate its problems, suggest ways and means for control, develop standards or codes of practice, furnish technical guidance, and conduct educational programs. It was believed that the health and safety of industrial workers could best be achieved by state and local health departments cooperating with employers and workers.

Technical problems in need of solutions included the need to develop more precise sampling instruments and analytical methods, better tools to detect cases of disease, increased engineering knowledge to control the factory environment, and more sophisticated methods to detect small amounts of toxic materials.

Along with funds, the Public Health Service offered leadership and guidance to the new state agencies to help them build industrial hygiene programs. The Division of Industrial Hygiene assigned the task of organizing the state agencies to Jack J. Bloomfield, who was instrumental in organizing the National Conference of Governmental Industrial Hygienists, later called the American Conference of Governmental Industrial Hygienists, in 1938. Bloomfield intended the organization to be the mechanism to help the Public Health Service carry out its mission.[17]

Until the 1930s industrial hygiene focused mainly on industrial medicine and was practiced by physicians. In the 1930s industrial hygiene began to take on new meaning. It began to include the nonmedical contributions of the physical sciences and engineering, thus emphasizing the environmental side of industrial hygiene.

By the early 1940s New Deal social legislation began to change the relationship between the federal government and the American people. One of the areas in which this occurred was occupational health.

In 1938 a number of individuals engaged in occupational health activities, in government agencies, formed the National Conference of Governmental Industrial Hygienists. Their objective was: "to promote industrial hygiene activities in all its aspects and phases, and to coordinate such activities in official federal, state, local and territorial organizations."[18] In the following year the American Industrial Hygiene Association and the American Occupational Health Nurses Association were organized. In 1946 the Academy of Occupational Medicine was founded. That year the United Mine Workers established labor's first major health and welfare fund.

During the 1930s and 1940s, medical responsibilities in the larger corporations began to extend beyond surgical treatment of injured workers. World War II stimulated growing public awareness of the need for occupational health, based on the perception that the productive capacity of the nation somehow depended upon the health and well-being of working men and women. Entry into World War II brought with it an urgency to produce. The outcome of the war depended as much on America's productive capacity as on the ability of its soldiers to fight and win on the battlefield. War mobilization stimulated a revolution in productivity.[19] It also stimulated occupational health activities.

Along with increased production came increased risk to the health and safety of working men and women and the necessity to curtail accidents and health hazards that could slow down production. Startling figures attest to the need for occupational health activities. Occupational casualties from Pearl Harbor December 7, 1941 to January 1, 1944 (two years) included 37,600 workers killed, which is 7,500 more than the military dead; 210,000 were permanently disabled and 4.5 million temporarily disabled, 60 times more than the military wounded and missing. According to the Office of Labor Production of the War Production Board, accidents caused approximately ten times as many lost labor hours as strikes; and deaths and injuries occurred on the job at a rate of 270 million lost employee days a year. Furthermore, estimates of losses sustained by labor, management, and the war effort because of disease and illness placed figures at 400 million man days, with an associated financial loss of $2 billion.[20]

Because hazards to the health and safety of workers were considered to be among the chief causes of low labor productivity, the Office of Labor Production suggested improvement of unsanitary and unsafe working conditions. The War Production Board established the Industrial Hygiene and Safety Section in the Office of Labor Production to improve health and safety on the job.

The federal government, faced with shortages of labor and materials and high rates of accident and illness, reasoned that it needed to define and carry out a policy to keep workers healthy, safe, and productive. This policy directly affected the growth of occupational health in the United States. But during the war occupational health did not become institutionalized in the federal government. The state industrial hygiene units and local agencies still retained responsibility for most occupational health activities.

In spite of limited resources, low funding, and lack of trained personnel, the war had stimulated new occupational health activities on both federal and state levels. In 1943, 47 state and local units were allotted $1,006,000, only $75,000 more than 1942. The 47 units had a total of 300 professional and clerical personnel. Of the 38 state organizations, 21 had staffs consisting of fewer than five professional personnel (including personnel loaned from the Division of Industrial Hygiene).[21]

After the war, the newly increased interest in occupational health began to shrink. State efforts lacked enforcement and also concordance between those charged with production and those charged with health and safety. Occupational health often amounted to little more than lip service or a slogan. Those to be protected had little or no input, and public interest and understanding dwindled, leading to retrenchment of the occupational health endeavor. The war had for only a short time focused interest on occupational health.

In a 1949 article in *Public Health Reports,* Victoria Trasco used the annual reports of the 45 industrial hygiene agencies in 38 states to describe governmental industrial hygiene activities for the fiscal year 1947–48. Fifty-eight state and local industrial hygiene units functioned on a "full or limited" basis in 44 state health departments, 2 state labor departments, 8 local health departments, the District of Columbia, and the territories of Hawaii, Puerto Rico, and Alaska. Appropriations totaled $2,500,723, of which 53.5 percent derived from federal sources and 46.5 percent from state and local sources. Agencies employed approximately 400 professionals, of which 66 percent were engineers and chemists, 11 percent sanitarians and inspectors, 10 percent physicians, 10 percent nursing consultants, and 3 percent a miscellaneous group of professionals and technicians.[22]

World War II presented an opportunity to develop occupational health activities in state and federal agencies, and by the end of the war a network of units had been established in nearly every state and in many large industrial cities. The problem was to keep the public aware and interested in occupational health and to have them accept it as a necessary part of industrial activity. Unfortunately, during the 1950s occupational health was not a major concern. Most Americans in the 1950s were uninterested in occupational health. Other public health problems, such as delivery of health care, and political issues dominated their thinking. The cold war and the fear of communism kindled by Senator Joseph McCarthy diverted American interest from issues such as occupational health.

Ironically, at the same time, air pollution, radiation hazards, and a growing chemical industry created new environmental risks in and outside of industrial plants. The manufacture of new consumer goods (synthetic fibers, plastics, detergents, and pesticides) caused environmental concerns and contaminated air in factories as well as outside factories. Although federal regulation of air and water pollution had already begun in the 1960s, it would take until 1970 to pass the Occupational Safety and Health Act. By then, environmental issues had already become public issues.

THE OCCUPATIONAL SAFETY AND HEALTH ACT, 1970

Safety and health laws, which had historically been left to the states, were piecemeal, varied in quality, and often unenforced. Every state had workmen's compensation laws and many had some type of health and safety law, but this patchwork pattern of coverage by a jumble of state and federal agencies proved inadequate. In fact, increasing injuries on the job in the 1950s and 1960s indicated the need to move beyond workmen's compensation laws and individual state responsibility toward a more comprehensive national policy of preventing accidents and unhealthy working conditions. The complexity of American industry, with its use of sophisticated processes and new chemicals, pointed to the need for research, nationwide statistical reports, and enforceable regulations to protect American workers. By the 1970s it was clear that these needs could be met only by federal legislation.

The movement for federal occupational safety and health legislation coincided with the environmental movement of the 1960s. As early as 1958, however, the federal government began to take a more active role in occupational health when Congress gave the Labor Department authority to set safety and health standards for longshoremen (The Longshoremen's and Harbor Worker's Compensation Act).[23]

In 1965 the U.S. Public Health Service report "Protecting the Health of Eighty Million Americans" outlined the new technological changes Americans faced, stated that old problems were far from solved, and called for a major national occupational safety and health focus centered in the Public Health Service. Following study of the Division of Occupational Health and the Public Health Service, a plan for reorganization of the Public Health Service was presented to make the service more responsive to the country's occupational health problems.[24] The Division of Occupational Health in the Public Health Service was apparently not functioning effectively, and its role in protecting worker health needed clearer definition.

In 1968 President Lyndon B. Johnson proposed a comprehensive occupational safety and health program to protect American workers on the job. He requested legislation to strengthen the authority and resources of the Secretary of Health, Education, and Welfare to develop an extensive research program that would be the basis for setting health and safety standards. The Secretary of Labor would set and enforce standards and impose sanctions that included civil and criminal penalties. That bill was never passed. A new bill was sent to Congress in 1969.[25]

Statistics and testimony at the extensive committee hearings for federal occupational safety and health legislation showed an increasing pattern of injury and illness on the job. Americans learned from these hearings that many factories and mines still remained perilous places. Testimony indicated that although the relationship between incidence of occupational disease and work in the asbestos, lead, coal mining, cotton, textile, and pesticides industries, to name a few, was well known, and although enough knowledge existed to control the work environment to prevent industrial diseases, the knowledge was not yet utilized for effective preventive measures. Statistics presented at committee hearings indicated increasing injuries on the job.[26, 26A] Information presented at the hearings illustrated the need for more than workmen's compensation laws and individual state responsibility for worker health and safety. It indicated the need for a comprehensive national policy to prevent workplace injuries and illness.

Congress had not embarked on completely uncharted waters in 1969. State efforts and some federal efforts had been around since the 1930s. By 1970, however, the consensus was that protecting workers could be achieved only by federal regulation and assumption of responsibility for worker health and safety. The undertaking was gigantic. Legislation covered more workers than had state laws, and the Department of Labor would have far-reaching regulatory, enforcement, and administrative responsibilities.

In 1970 Congress passed the Occupational Safety and Health Act, a watershed in the history of occupational health in the United States. It created OSHA (Occupational Safety and Health Administration), a regulatory agency, to set and enforce regulations to control and prevent safety and health hazards. It is in the Department of Labor and headed by an Assistant Secretary of Labor for occupational safety and health appointed by the President. OSHA's functions include setting mandatory safety and health standards, inspecting workplaces to ensure compliance with standards, proposing penalties for violators, and providing education for workers, employers, and the public. The act also created NIOSH (National Institute for Occupational Safety and Health), a research agency, and part of the Centers for Disease Control of the United States Public Health Service in the Department of Health and Human Services (formerly the Department of Health, Education, and Welfare). The director of NIOSH is appointed by the Secretary of Health and Human Services. NIOSH functions include conducting research and related activities to develop criteria, recommending these criteria to OSHA for standard setting, and providing professional education and dissemination of health and safety information.

It is extremely difficult to assess OSHA's impact on injury rates and exposure levels. Research can be divided into two groups: studies that find a statistically significant, but small effect and those that do not find any. An Office of Technology Assessment report stated that "limited research on exposure levels appears to show positive effects for hazards that were the subjects of new or revised OSHA regulations during the 1970s—vinyl chloride, asbestos, cotton dust, lead."[27]

Nevertheless, the Occupational Safety and Health Act is universally considered a landmark in the history of labor and public health legislation. The impact of the act on working men and women, with its purpose "to assure so far as possible every working man and woman in the Nation safe and healthful working conditions," is difficult to overestimate. Basic to the act is the idea that in-plant health and safety is the responsibility of the employer.

The impact of the Occupational Safety and Health Act upon occupational health professionals (occupational health physicians, industrial hygienists, and occupational health nurses) is also enormous. Federal regulations changed the way these professionals functioned.

> Prior to its passage, there was little need for documentation of our work or employer's efforts. The creation of duties by the OSHAct requires that records be kept to demonstrate that employers have discharged their duties. Our reports are not only read they are scrutinized. There are legal contests revolving around our efforts, addressing the accuracy, conclusions and recommendations emanating therefrom, as well as the appropriateness of the responses of those line managers compelled to act. The law has specifically defined duties and responsibilities as they were never defined in the past.[28]

The number of occupational health professionals has grown rapidly since 1970.

Perhaps the act's greatest impact has been that it sensitized and educated the work force and the general public to the hazards of work. The extent to which it stimulated the occupational health professions is re-markable, but even more remarkable is the extent to which it made the American public aware of the relationship between work and health, made occupational safety and health a public issue, and increased labor union activity and concern. The act and subsequent labor activity paved the way for right-to-know laws.

Historically the Occupational Safety and Health Act can be viewed as the culmination of 60 or more years of effort toward a safe and hazard-free workplace. Creation of a federal agency responsible for occupational health with inspection and enforcement authority took more than half a century. Looking backward, a pattern of increasing federal involvement and assumption of responsibility can be discerned.

References
1. Derickson, Alan. *Workers' Health, Workers' Democracy: The Western Miners' Struggle 1891–1925*. Ithaca: Cornell University Press, 1988.
1A. Corn, Jacqueline. *Environment and Health In Nineteenth Century America: Two Case Studies*. New York: Peter Lang, 1989.
1B. Corn, Jacqueline. *Protecting the Health of Workers*. Cincinnati: ACGIH, 1990.
1C. Rosner, David and Gerald Markowitz. *Deadly Dust: Silicosis and the Politics of Occupational Disease in Twentieth Century America*. Princeton: Princeton University Press, 1991.
1D. Rosner, David and Gerald Markowitz, eds. *Dying for Work: Workers Safety and Health in Twentieth Century America*. Bloomington: Indiana University Press, 1987.
1E. Sellers, C. "The Public Health Service's Office of Industrial Hygiene and the Transformation of Industrial Medicine." *Bulletin of the History of Medicine* 65(1): 42–73.
1F. Rosner, D. and G. Markowitz. "A Gift of God: The Public Health Controversy Over Leaded Gasoline During the 1920s." *American Journal of Public Health* 75(4): 344–352.
1G. Fox, D. and J. Stone. "Black Lung: Miners' Militancy and Medical Uncertainty, 1968–1972." *Bulletin of the History of Medicine* 54: 43–63.
2. Corn, Jacqueline. *Protecting the Health of Workers*. Cincinnati: ACGIH, 1989.
3. Hoffman, Frederick. *Industrial Accident Statistics*, Bulletin 157 Washington: U.S. Department of Labor, Bureau of Labor Statistics, 1915 p 5.
4. MacLaury, Judson ed. *Protecting People at Work*. Washington: Government Printing Office, 1980.,
5. MacLaury, Judson. "The Job Safety Law of 1970: Its Passage Was Perilous." *Month Labor Review*, March 1981, 19.
6. Eastman, Crystal. *Work Accidents and the Law*. New York: Russell Sage Foundation, 1916.

7. Ibid., 13.
8. Hamilton, Alice. *Exploring the Dangerous Trades*. Boston: Little, Brown & Co., 1943.
9. Ibid., 115.
10. MacLaury, "Job Safety Law," 19–20.
11. Hoffman, *Industrial Accident Statistics*.
12. Ash, Shalom. "The Triangle Fire". In *Working Women*, edited by N. Hoffman and F. Howe. New York: McGraw-Hill, Feminist Press, 1979.
13. Corn, Jacqueline. "United States Occupational Safety and Health in the 1930s: The Formative Years of the American Conference of Governmental Industrial Hygienists." *Applied Industrial Hygiene*. 3(4): F8–11.
14. MacLaury, Judson. "The Division of Labor Standards: Laying the Groundwork for OSHA." *Applied Industrial Hygiene* 3(12): F8.
15. Williams, R. C. *The United States Public Health Service*. Washington: Commissioned Officers Association of the USPHS, 1951.
16. Bloomfield, J. J. "Development of Industrial Hygiene in the United States." *American Journal of Public Health* 28(12): 1392.
17. Corn, *Protecting Health of Workers*, 14.
18. Transactions of the First Annual Conference of Governmental Industrial Hygienists, Washington, D.C., 27-29 June 1938.
19. Corn, Jacqueline. "Historical Perspective on Work, Health and Productivity." In *Work, Health and Productivity*, edited by G. Green and F. Baker. Oxford University Press, forthcoming.
20. Transactions of the Seventh Annual Conference of Governmental Industrial Hygienists, St. Louis, May 1944.
21. Bloomfield, J. J. "Governmental Industrial Hygiene Practice in 1942." *Industrial Medicine* July 1943.
22. Trasco, Victoria. "The Work of State and Local Industrial Hygiene Agencies," *Public Health Reports* 64(15): 471.
23. MacLaury, "Job Safety Law," 20.
24. "Protecting the Health of Eighty Million Workers." In Transactions of the Twenty-Eighth Annual Meeting of the American Conference of Governmental Industrial Hygienists, Pittsburgh, 15–17 May 1966.
25. Mintz, B. *OSHA: History, Law and Policy*. Washington: The Bureau of National Affairs, Inc., 1984.
26. Hearings before the Subcommittee on Labor, Committee on Labor and Public Welfare, U.S. Senate, *Occupational Safety and Health Act of 1968*, 90th Cong., S2864, 15 February, 12, 19, 24, 28 June, and 2 July 1968. Washington: U.S. Government Printing Office, 1968.
26A. Hearings before the Subcommittee on Labor, Committee on Labor and Public Welfare, U.S. Senate, *Occupational Safety and Health Act of 1970*, 91st Cong., S2193 and S2788, 30 September, 4, 21, 24, 26 November 1969; 7 March, 10, 28 April 1970.
27. *Preventing Illness and Injury in the Workplace*. OTA-H-256 Washington: U.S. Congress, Office of Technology Assessment, 1985.
28. Corn, M. The Progression of Industrial Hygiene. *Applied Industrial Hygiene* 4(6): 155.

Chapter 2

Evolution of Mandatory Standards

Modern standard setting to minimize the exposure of workers to harmful chemicals and physical hazards involves technical as well as social choices and decisions. As a result, it often transcends science and spills into public debate to become a source of scientific and political controversy. When debated scientific information is set in the context of vested interests, it is further complicated by political relationships. In this context, setting standards becomes a political struggle that only intensifies when it occurs between those who view standard setting as either primarily a technical or a political process.

The concept of occupational standards, as we understand it today, is the philosophical basis for creating a healthy workplace. It is a recent idea associated with achieving an acceptable level and duration of exposure to a potentially toxic agent, based upon evidence that assumes predictive validity for health effects that will follow if the standard is exceeded. A standard implies documentation of the link between exposure and effect. Thus measurement is essential for setting standards.

A *regulation,* in contrast to a standard, is a promulgation by a body authorized to enforce enabling legislation. Regulations frame, in legal context and in specific terms, requirements imposed by an agency on regulatees; they usually serve as an umbrella for standards.[1]

This chapter focuses on three historical aspects of standard setting. They have been selected because each one is significant in the history of the development of procedures for standard setting in a technological society. First, developing concepts of environmental measurement are explored. The concept of environmental measurement or exposure assessment is the scientific underpinning utilized for the modern standard-setting process. Gaps in this knowledge base often contribute to conflict and controversy. In exposure to a potential toxic or physical agent, the effect is proportional to the amount the individual receives. Thus, it is important to measure the extent of exposure in a standardized, repeatable manner. The evolution of methods to measure health risks is often a slow and difficult process. For regulatory purposes, measurement must be performed in a standard manner with all concerned utilizing similar techniques.

Second, the voluntary guidelines for occupational exposure limits established by the American Conference of Governmental Industrial Hygienists Threshold Limit Values (TLVs) are placed in historical perspective. In the early part of the twentieth century, occupational health professionals (industrial hygienists, including physicians and toxicologists) formed associations to pool their relevant knowledge about hazards and approaches to reducing health and safety risks in the workplace. The formulation of this knowledge often led to a voluntary set of guidelines for action based on an occupational health conceptual framework or set of principles, for example, dose-response. One very influential set of voluntary guidelines is the TLVs.

Third, the process of setting mandatory legal occupational exposure standards, set in motion by the Occupational Safety and Health Act of 1970, is discussed. It represents the government's assumption of responsibility for uniform safety and health requirements in the workplace. It is the culmination of efforts of professionals, workers, and employers to deal with hazards and explicitly recognizes that government effort is required to achieve effective results. Mandatory standards are the major tools utilized in that effort. Implicit in this approach is the right of working men and women to a safe and healthful environment. That right supersedes any earlier concepts of occupational risks as inherent in an employee's agreement to perform work in return for compensation.

PUBLIC HEALTH AND STANDARDS

There is a long history of standard setting to protect public health. Laws have existed to improve public health and the environment since ancient times. As societies grew more complex, laws became more elaborate. Major changes in social philosophy and technology guided approaches and utilization of legal mechanisms to improve public health. Urban growth and industrialization required individuals to develop regulations and standards to protect themselves. The degree to which science was incorporated into the law is reflected by the specificity of the law. In the ancient world, for example, when public health officials dealt with public baths or food supplies, they attempted to regulate environmental hazards to improve public health without measurement techniques; they utilized observation and experience.[2]

In general, the basis for preventive public health law in the United States was set in the nineteenth century. Preventive measures for sanitation and for contamination of food to protect the general welfare usually resided in the police power of the state or local board or department of health. The police power usually assigned regulatory authority to an administrative body that passed specific regulations and/or standards to achieve the purpose of the enabling legislation.[3] Prior to the twentieth century, few public health laws existed to protect men and women in their workplaces.

DEVELOPING CONCEPTS OF ENVIRONMENTAL MEASUREMENT

Exposure assessment can be traced back to early approaches to measuring the extent of hazards in the workplace. In order to understand standard setting in twentieth-century America, exposure assessment must be explored. It is a critical concept in decision making to regulate the workplace and one that has been evolving since the beginning of the twentieth century. Developing concepts of measurement and assessment impacted on decisions to control occupational hazards. Based on the assumption that measurement and action to control a dangerous substance are intertwined, it is impossible to discuss one without the other.

The history of asbestos measurement is utilized here to illustrate both changing concepts of measurement and assessment and the United States experience in setting acceptable workplace exposure levels to control risk from a toxic material.

Asbestos has probably received more attention in the United States from practitioners and investigators in the fields of industrial hygiene, occupational medicine, occupational epidemiology, and toxicology than any other toxic agent. It has created economic crises in major corporations responsible for its production and distribution and caused two major corporations to enter into Chapter 11 bankruptcy. Concern related to health hazards associated with asbestos ranges from those occupationally exposed to those experiencing nonoccupational exposure. Asbestos concerns precipitated regulation in the occupational environment (OSHA) and the nonoccupational environment (Environmental Protection Agency [EPA] and Consumer Product Safety Agency [CPSA]). Perhaps more is known about asbestos toxicity and the dose-response curve than about any other industrial material. Today, while there is still concern that people who work with asbestos are not adequately protected, anxiety and fear are mounting over the longer-term effects of exposure to asbestos in the general environment.

The criteria used to measure risks associated with asbestos have changed since the 1930s, based on proliferation of asbestos utilization, changes in understanding of the diseases caused by asbestos, ability to make more accurate and sensitive measurements, public knowledge, and federal regulation of the workplace in 1970, which called for legally mandated standards.

As noted earlier, standard setting is a policy decision closely linked to technical and scientific knowledge. The history of setting standards for asbestos, a toxic substance with a toxicological database probably more extensive than that for any other toxic substance, demonstrates that there is always an incomplete database to answer the associated technological and scientific questions, such as magnitude of biological effects and efficiency of controls. The incomplete database often causes political controversy.

In the 1960s and 1970s, concerns about asbestos hazards centered on workers who had been exposed to large amounts of fibers at work because of intense utilization, minimum controls, and minor precautions on the job. Manifestations of asbestos diseases appeared with increased frequency in shipyard workers and others who had been exposed to massive amounts of asbestos fibers on the job. Because of the long latency period associated with asbestos-induced cancer, it took until the 1960s to fully appreciate the legacy of death and disease that resulted from earlier high exposures to asbestos. At that time Irving J. Selikoff, Jacob Churg, and E. Cuyler Hammond presented epidemiological evidence that insulation workers who had dealt with asbestos for 20 years or more were dying of cancer and complications of asbestosis at alarming rates.[4] In 1972 asbestos became the first toxic material regulated by the new Occupational Safety and Health Administration.

Although asbestos had been used in small quantities for centuries, large-scale asbestos mining and commercial production started in the twentieth century and greatly accelerated during World War II. Asbestos consumption rose from under 100,000 tons in 1912 to approximately 750,000 tons during World War II and 800,000 in the 1970s.[5]

Industries that manufactured asbestos products or utilized them employed millions of people. In the mid-1970s it was estimated that more than 37,000 persons were employed in manufacture of primary asbestos products, 300,000 worked in secondary asbestos industries, and millions in asbestos-using consumer industries, including 185,000 in shipyards and almost 2 million in automotive sales, service, and repair.[6]

At the present time it is universally accepted that asbestos exposure causes serious illness and death and that the major pathological effects of asbestos result from inhalation of fibers suspended in ambient air.

The health effects of asbestos include asbestosis, a chronic, restrictive lung disease caused by inhalation of asbestos fibers, the first known disease associated with exposure to asbestos and associated with heavy occupational exposure to asbestos.[7] Lung cancer is not specifically associated with asbestos exposure as are asbestosis and mesothelioma because lung cancer also has a history of association with cigarette smoking. Although it is recognized that asbestos in the absence of cigarette smoking can induce lung cancer, issues of causation are often raised when lung cancer develops in asbestos workers who smoke.[8] Mesothelioma, a rare cancer of the surface-lining cells of the pleura (lung) or peritoneum (abdomen), generally spreads rapidly and over large surfaces of either the thoracic or abdominal cavities. No effective treatment exists for mesothelioma. It occurs among insulators, those who work in asbestos plants, and shipyard workers. Mesothelioma can also occur among persons living in the same house as asbestos workers or near asbestos mining and milling. Mesothelioma, like asbestosis, is specifically linked to asbestos exposure.[9]

Documentation of cases of asbestos-related disease began early in the twentieth century.[10, 10A] At that time, lack of knowledge about the risks associated with asbestos and the small amounts of mineral used limited the observations and understanding about the relationship between asbestos and disease. In fact, the fibrotic disease described by Murray, Auribault, and others at the beginning of the twentieth century did not receive a name until W. E. Cooke called it *asbestosis* in 1927.[11]

As asbestos production began to expand, observations of asbestosis increased. In 1928 and 1929, the British government undertook an investigation of the condition of asbestos textile factory workers and reported to Parliament in 1930 that "inhalation of asbestos dust over a period of years results in the development of a serious type of fibrosis of the lungs." The commissioners recommended dust suppression. British Asbestos Industry Regulations followed in 1931.[12] Clinical reports in the United States also confirmed occurrence of asbestosis among asbestos workers.[13, 13A, 13B] A number of books on public health, medicine, and related subjects began to incorporate sections on industrial hazards associated with asbestos dust.[14, 14A–14D]

By the 1940s asbestos dust was identified as dangerous and unhealthy; inhalation of the dust over a period of years could cause asbestosis. But a 20-year gap existed between the first identification of asbestos as the cause of fibrotic disease and general acceptance of asbestos dust as a health hazard. Asbestosis is also coincident with high levels of dust exposure, and in the early years of the twentieth century workers were exposed to heavy concentrations of asbestos dust. It would take years before significant reduction of dust occurred. In the meantime American and European industries found new applications and used more and more asbestos, seldom considering the health of workers.

The first reported case of association of asbestos dust and lung cancer appeared in 1935,[15] but the first rigorous epidemiological study appeared in 1955, when Sir Richard Doll documented that the risk of lung cancer was increased tenfold among a group of men employed 20 or more years at an asbestos textile plant in northern England.[16] To complicate the matter, the latency period for cancer was longer than for asbestosis, and lower levels of dust could cause cancer. Nevertheless, the foreshadowing of the tragedy to come had already begun in the 1930s, when British and American medical reports associated asbestos exposure with the development of lung cancer. In 1964 the landmark Conference on Biological Effects of Asbestos, organized by the New York Academy of Sciences, resulted in a consensus among investigators of different nations that asbestos was a cause of lung cancer.[17]

The lengthy route from initial cognizance to confirmation and acceptance of the association between asbestos and cancer once again, as in the case of asbestosis, took decades. Tragically, large numbers of workers in the meantime became desperately ill, and many of them died from a painful disease. Furthermore, in the 1960s it was becoming clear that the risk of disease was not confined solely to workers in mining and manufacturing but that it extended to shipyard workers, insulation workers, and many others outside the primary or fixed-place industries.

In the United States, minimal attempts to control occupational exposures to asbestos began in the 1930s. A body of technical and medical literature about fibrotic disease associated with asbestos existed. Part of the emerging new literature and technical information included measurement or assessment of asbestos dust and the development of the concept of dose-response.

In the 1930s industrial hygiene and occupational medicine professionals developed the concept of a dose-response relationship that implied measurement and assumed a linear relationship between dose and response. Since then, two essential principles of in-plant safety have been utilized. First, a systematic dose-response relationship exists between the severity of exposure to the hazard and the degree of response in the population exposed. As the level of exposure decreases, there is a gradual corresponding decrease in the risk of injury. Second, the risk becomes negligible when the exposure falls below certain acceptable levels. Analysis of environmental factors is implied in the dose-response concept.

Prior to the utilization of exposure guidelines and techniques for measuring hazards in the workplace, observation and experience were the keys to controlling hazards. In 1948 Teleky wrote,

Twenty years ago [1928] the only method in use was a periodic examination of workers. To this has been added measurement of the amount of harmful substances in the air and the determination of their effect on health. For the development of industrial hygiene, it was necessary to find ways of measuring noxious substances in the air. It was not sufficient to rely on general terms like "much" or "damaging" or to infer from the incidence of disease in the factory that the air contained a damaging amount of dust or other substances. Only in the last decade has a fairly exact determination been possible, although attempts have been made since the middle of the last century.[18]

Early attempts to measure dust content in air consisted of merely allowing dust to settle or drawing a measured amount of air through a filter by suction. These simple instruments utilized cotton, asbestos, or nitrocellulose as filters and consisted of two bottles of 25-liter capacity each. One bottle, filled with water, stood on a higher level and the other on a lower level. The water ran from the upper to the lower bottle through a pipe that had a piece for aspiration. The apparatus was very cumbersome.

Another dust-measuring device constructed on different principles utilized a method that impinged a jet of dusty air on a sticky substance. Kotze's Konimeter, constructed to measure dust in mines, allowed dusty air to impinge at a high velocity through a narrow nozzle against a plate coated with an adhesive substance, thus retaining the dust particles for subsequent microscopic examination on the plate.[19]

In 1922 Leonard Greenburg and G. W. Smith, both employed by the United States Bureau of Mines, combined the principle of collecting dust by impingement with the water-washing or bubbling method to construct the impinger.[20] The impinger method of dust collection remained the standard dust collection method for over forty years. The dust was counted after dust-sampling apparatus took dust out of a measured amount of air. Dust particles collected by the impinger method were counted by means of a microscope and later through projection and microphotography.

The corollary to measuring airborne hazards and to analysis of environmental factors in the workplace was the need to determine the concentration of a material that would not actually cause injury, in other words, to determine concentrations at which individuals were exposed but not injured. A body of data accumulated that gave exposures not associated with injurious effects for a variety of compounds. Determining "safe" concentrations was and still is a controversial and value-laden concept, in part because of the uncertainties associated with the always incomplete database. Terms for measurement of what was to be considered acceptable include: MAC (Maximum Allowable Concentrations), PEL (Permissible Exposure Limits), and TLV (Threshold Limit Values).

In the United States industrial hygienists began to publish tables of MACs, that is, the upper limit of concentration of an atmospheric contaminant that they believed would not cause injury to an individual exposed continuously during the working day and for indefinite periods of time.[21]

The first suggested guidelines for asbestos appeared in 1938 in the Dreessen study, *A Study of Asbestosis in the Asbestos Textile Industry,* prepared for the asbestos textile industry in North Carolina and published as a Public Health Bulletin.[22] The authors studied the process that produced dust in the factories and recommended dust control. Dreessen and associates made dust counts with a midget impinger, estimated the dust exposure of each worker, and concluded,

> The percentage of persons in different occupational groups who were affected by asbestosis or any of its symptoms varied with the average dust concentration to which they were subjected and with their length of employment. The only cases of asbestosis, three in number, found below 5 million particles per cubic foot were diagnosed as doubtful; well established cases occurred at higher concentrations. It appears from these data that if asbestos dust concentrations in the air breathed are kept below this limit new cases of asbestosis would not appear.[23]

Furthermore, the study's authors said, "Because clean-cut cases of asbestosis were found only in dust concentrations exceeding 5 million particles per cubic foot, and because they were not found at lower concentrations, 5 million particles per cubic foot may be regarded tentatively as the threshold value for asbestos dust exposure until better data are available."[24] Sampling methodology for airborne fiber concentration utilizing the impinger involved counting all particles present.

Today the Dreessen study is regarded as a flawed and extremely limited cross-sectional epidemiological investigation. Nevertheless, the Dreessen numbers for "safe" levels of asbestos in air stood for the next 30 years and became the basis for the TLV guideline. That guideline was not critically appraised until the 1964 meeting of the New York Academy of Sciences that led to publication of *Biological Effects of Asbestos,* which confirmed asbestos as a carcinogen.

The first edition of *Documentation of Threshold Limit Values* (published by the American Conference of Governmental Industrial Hygienists [ACGIH] in 1962) stated,

> The present threshold limit relates to the prevention of asbestosis. It was recommended by Dreessen, et al., after study of 541 employees in three asbestos textile plants using Chrysotile. Only three doubtful cases of pneumoconiosis were found in those exposed to dust concentrations under 5 mppcf, whereas numerous well marked cases were found above 5 mppcf. Counts were from impinger-collected samples in ethyl alcohol and distilled water. Both fibrous and non-fibrous particles were counted, but the latter greatly predominated. While chemical analyses of collected samples of airborne dust corresponded to those of settled dust samples it is believed that dust counts of particulates by conventional methods can be expected to give only an indirect measure of the risk of asbestosis because of the great relative importance of long fibers.[25]

The impinger used in the Dreessen study was developed to measure particles and fibers in air. Because asbestosis was a fibrotic disease, measurements were developed to quantitate the risk of the environment to cause fibrotic disease. But in the 1960s there was growing concern because of the increasing realization of the carcinogenic properties of inhaled asbestos. With new understanding that asbestos forms only fibers in air, due to mineral breakage characteristics, measurements began to focus on fibers and not on fibers and particles. New measurement techniques were developed, and fibers per cubic centimeter of air were counted rather than fibers and particles. Increased knowledge of which fibers reached the lung pointed to the need for new ways to measure and evaluate samples. Furthermore, because carcinogens are assumed to have no threshold, the concept of Threshold Limit Value needed reevaluation in the case of asbestos. Table 2-1 indicates the changes in asbestos standards from 1938 to 1986. It illustrates changes that occurred due to new laws (OSHA) as well as to increased technical understanding.

In the 1971 third edition of the ACGIH's *Documentation of Threshold Limit Values*, there is a more critical paragraph on the Dreessen study, as well as a lengthier description of results relevant to a standard. It treated the Dreessen study as follows:

> A conference on the biological effects of asbestos in 1965 called attention to the very real probability that the 5 mppcf limit recommended by Dreessen is inadequate to give complete working-lifetime protection against all forms of asbestos. Medical data on which the limits had been based were inadequate; more than half of the asbestos workers studied were under 30 years of age

TABLE 2-1.   U.S. Asbestos Standards

Year	Sponsor	Status	Million Particles/cm	Fibers/cc
1938	Dreessen et al.	Recommended TLV	5	30[a]
1946	ACGIH	Adopted	5	30[a]
1970		Adopted	2	12[a]
1971		Proposed	—	5
1971	OSHA	Emergency TWA[b]	—	2
1975		Proposed	—	0.5
1976		Adopted	—	2
1976	NIOSH	Recommended	—	0.1
1983	OSHA	ETS[c]	—	0.5
1984	OSHA	Proposed	—	0.4 or 0.2
1986	OSHA	Adopted	—	0.2

[a] Approximate fiber equivalent.
[b] TWA = Time Weighted Average
[c] ETS = Emergency Temporary Standard

and thus provided an insufficient exposure time for asbestosis to develop. Of the 105 workers exposed to less than 5 mppcf, 82 had worked less than 5 years; 101 less than 10 years; only 4 had more than 10 years exposure. Seven of 36 workers exposed to 5–9.9 mppcf had asbestosis; 3 of 50 workers exposed to 10–19.9 mppcf for less than 5 years had asbestosis. Moreover, it was a "point-in-time" study; many of the ill were missing and the dead uncounted, hence not considered in the overall evaluation of the limit.[26]

Until the passage of the Occupational Safety and Health Act of 1970, the TLV for asbestos remained merely a recommended standard, not an enforceable one. Before the 1970s nobody seemed seriously concerned with their adequacy. The growing concern in the 1970s occurred in part because of the increasing realization of the carcinogenic properties of inhaled asbestos and the increasing utilization of asbestos. The TLV for asbestos was developed to protect against asbestosis, a fibrosis of the lungs. As the carcinogenic effects of asbestos became known, the TLV was proven inadequate to protect for cancer.

Sampling methodology for airborne asbestos has also changed since 1938. New instruments were devised and new measurement techniques developed. Industrial hygienists no longer counted the same types of dust particles, used the same methods, or asked the same questions. Medical scientists redefined diseases associated with asbestos. Nevertheless, the TLV or guideline based upon the Dreessen study remained on the books for more than 30 years, inadequately addressing the need to control asbestos to the lower levels related to carcinogenic effects.

OSHA has regulated asbestos since 1971. Under present regulatory procedures in the United States since 1970, OSHA's permanent health standards for airborne contaminants require that an initial determination be made of concentration of the agent in air. Until OSHA, there were no federally mandated exposure standards, except for those in regulations of the Walsh-Healey Public Contracts Acts of 1968 and a few states that established their own values. Initial promulgation of an OSHA standard in May 1971 was a 12 fibers per cubic centimeter (f/cc) permissible exposure limit (PEL). In December of 1971 OSHA issued an Emergency Temporary Standard of 5 f/cc and in June 1972 a new final standard of 5 f/cc PEL.[27] But these limits were intended primarily to protect workers against asbestosis and to offer only a limited degree of protection against an asbestos-induced form of cancer.

In October 1975 OSHA published a Proposed Rulemaking to revise the asbestos standard because the agency believed that "sufficient medical and scientific evidence had been accumulated to warrant the designation of asbestos as a human carcinogen" and that "advances in monitoring and protective technology made re-examination of the standard desirable."[28] The 1975 proposal was to reduce the PEL to 0.5 f/cc. The basis for the 1975 proposal's reduction in PEL to 0.5 f/cc was OSHA policy at the time for carcinogens that assumed no safe threshold level was demonstrable. The Occupational Safety and Health Act required OSHA to set the PEL at a level as low as technologically and economically feasible. (This policy was rejected by the Supreme Court in 1980 in the Benzene Decision, *IUD vs. API*.[29])

In 1976 the PEL was reduced to 2 f/cc. The 2 f/cc limit remained in effect until 1986. It took from 1976 to 1986 to further reduce the PEL to 0.2 f/cc. "The 0.2 f/cc 8 hour limit reduces significant risk from exposure and is considered by OSHA, based upon substantial evidence in the record as a whole, to be the lowest level feasible."[30]

In little more than a 50-year period approaches to assess environmental risk from asbestos have undergone profound change related to a number of factors, including (1) new technologies, (2) new measuring instruments, (3) new biological data, (4) increased utilization of asbestos, (5) regulatory imperatives, and (6) a redefinition of health risks associated with asbestos. These factors, in combination with social changes and public awareness of occupational diseases, caused reevaluation of asbestos policies including environmental assessment.

Early air-sampling instruments were developed to measure particles and fibers in air. Because asbestosis was a fibrotic disease, all measurements were developed by measuring fibers and particles. Subsequently, because of the understanding that asbestos forms only fibers in air, measurement began to focus on fibers and not fibers and particles. Increased knowledge of which fibers reached the lungs pointed, once again, to the need for new ways to sample and evaluate air. At the same time changes in understanding of the disease process led to the need for a better way to measure risk, including reevaluation of the concept of a threshold. In the 1960s acceptance of the knowledge that asbestos causes cancer as well as fibrosis raised the question of which fiber sizes should be measured and what the standard should be. In 1985 and 1986, the prevailing scientific assumption of a linear nonthreshold dose-response curve for carcinogens, applied to estimates of risk for asbestos exposure in the workplace, led OSHA to lower the United States standard.

This illustrates how closely linked the state of the art of scientific and technical knowledge is with policy decisions to utilize the concept of a standard to control a toxic substance in industry. In the best of all possible worlds, the science should permit valid estimates of risks and techniques for measurement that lead to the desired control. If awareness comes early, then the process of policy-making should be an iterative one with incrementally more restrictive control as knowledge of the toxicant improves. In the case of asbestos, as with many other toxic materials, the imperative for control was political, not scientific. The political decision to regulate the workplace environment set in motion utilization of the science that would permit valid estimates of risk and the techniques for measurement and control. Setting acceptable workplace exposure levels to control inhalation risk from asbestos did not occur on a wide scale until after 1970. On the other hand, after at least 50 years of less-than-intense development, the American approach to regulating hazards, including asbestos, in the workplace is now based on governmentally published explicit techniques for assessing risks that permit valid risk characterization.

TLV VOLUNTARY GUIDELINES

Early in the 1940s the ACGIH* began to develop guidelines for occupational exposure limits now known as TLVs. TLVs were based on analysis of environmental factors as a means to control toxic hazards in the workplace. The critical assumption in the TLV concept is that there is a level of exposure below which no adverse health effect occurs. TLVs were based on the concept of a dose-response relationship that implied measurement and employed two essential principles.

---

* The American Conference of Governmental Industrial Hygienists is a professional organization founded in 1938. It limits membership to government and university employees. It originally was called the National Conference of Governmental Industrial Hygienists.

1. A systematic dose-response relationship exists between the severity of exposure to a hazard and the degree of response in the population exposed. As the exposure level decreases, there is a gradual reduction in the risk of injury.
2. The risk becomes negligible when the exposure level falls below certain acceptable levels.

From the onset, establishing numerical values for occupational exposure limits acceptable to society was both a political and scientific activity. It was, and still is, value laden, highly controversial, and rife with conflicting principles and processes.

This section focuses on the early history of the TLVs, including how, why, and by whom they were developed and utilized and the relation among development, nature of utilization, and the recent events that highlighted the controversial and political nature of the TLVs.

The TLV approach sought to determine the concentration of material that already caused injury and to set threshold values based on the concentrations measured. It was believed that setting threshold values for the working environment permitted control of the environment and reduced the likelihood of injury to a minimum. Values needed to be periodically reviewed so that they would not be "frozen" by time and usage and so that newer, more accurate data would not be ignored.

Fear that industrial interest, including input into determining TLVs, and manipulation of the TLV approval process and procedures, appeared simultaneously with the first TLVs. How to determine threshold limits, who should be responsible for developing them, and how to enforce them once they were developed were issues associated with the early and subsequent development of Threshold Limit Values. TLVs, even though controversial, remained the only game in town until 1970, when the Occupational Safety and Health Act became law. After OSHA the ingredients of a federally mandated occupational health standard were legally defined and the promulgation process specified.

In 1940, in recognition of the need for exposure guidelines, NCGIH members discussed the merit of appointing a committee to consider threshold limits. Letters written by leaders of the organization reflect this discussion: "Most of the threshold limits are in a state of flux and the committee is urgently requested to sift the information and data available to provide the field forces with fairly uniform yardsticks."[31]

In 1941 the NCGIH divided its Committee on Technical Standards into two subcommittees, one on technical standards and the other on threshold limits. The Threshold Limits Subcommittee consisted of Manfred Bowditch, Philip Drinker, Lawrence Fairhall, Al Dooley, and William

Frederick (chairman). By 1944 the Threshold Limits Subcommittee had become a separate standing committee. Its members presented an extensive list of MAC* values to the conference for use during 1946, with the definite understanding that the list be subject to annual revision.[32] In 1946 the NCGIH published a "List of Threshold Limit Values" in the *Archives of Industrial Hygiene and Occupational Medicine* and distributed reprints to its members and others interested in the guidelines.

In 1948 the chairman of the Threshold Limits Committee commented on the philosophical basis for the Threshold Limits:

> The composition of this committee is not fixed but is changed by annual appointment and the Threshold Limit Values are similarly maintained in a fluid state of annual revision. For a number of years the committee has subjected these values to careful scrutiny and attempted to bring them into close conformance with practice. . . . In view of the annual revision of threshold limit values, it has been the purpose of the Conference to seek values which on the one hand, protect the individual workman, and on the other hand impose no impossible burden on the manufacturers. The balance is difficult to achieve, but can be attained by just such a series of adjustments as is provided for in the constitution of the conference. There is no industrial poison so potent, so virulent, that it cannot be manufactured under carefully controlled conditions.[33]

In 1948 criticism of TLVs included that (1) values came from consulting the membership at large, (2) members were not consulted, (3) publication of values in the *Industrial Hygiene Newsletter* made it appear that the values were those of the Public Health Service, and (4) values arrived at were not based on scientific evidence.[34] The following disclaimers appeared as part of what might be called permanent or standard Maximum Allowable Concentration values:

> It must be borne in mind that all our values at the present time are fluid and subject to annual revision. They should not be adopted as fixed or legal values, but merely as guides to assist us in defining more or less safe working conditions. . . . People vary greatly in response to drugs and toxic substances. Therefore it is a figment of one's imagination to think that we can set down a precise limit below which there is complete safety and immediately above which there may be a high percentage of cases of poisoning among the exposed.[35]

---

* TLVs were first called *maximum allowable concentrations* (MACs). The first American list of chemical exposures was prepared by Manfred Bowditch, Cecil Drinker, Philip Drinker, H. H. Haggard, and Alice Hamilton.

The use of Threshold Limits was discussed at every ACGIH meeting in the 1950s. The introduction to the published tables explained their intended uses. For example, in 1953 the Committee on Threshold Limits prefaced the publication of threshold values with

> These values are based on the best available information from industrial experience, from experimental studies and when possible, from the combination of both. They are not fixed values, but are reviewed annually by the Committee on Threshold Limits for changes, revisions, or additions as further information becomes available. Threshold limits should be used as guides in the control of health hazards and should not be regarded as fine lines between safe and dangerous concentrations.[36]

In spite of statements that threshold limits were not designed for and should not be used or modified for community air pollution control programs, state regulations often included TLVs by incorporating them into codes and laws.[37] The problems associated with improperly utilized TLVs continued, and the committee kept revising the introduction in hopes that the revisions would lead to a better understanding of the meaning and application of threshold limits. In 1959 the ACGIH again reaffirmed the principle that threshold limits should be used as guides in the control of health hazards and not be regarded as fine lines between safe and dangerous concentrations.[38]

During the 1950s information developed by the Committee on Threshold Limits received wide circulation. The list was published in the *Archives of the American Medical Association,* reprinted in *The Industrial Hygiene Digest* and *The Guide of the American Society of Heating and Air Conditioning Engineers,* circulated by the Mine Safety Appliances Company, and mailed to all members of the ACGIH and all state industrial hygiene units.

Although controversial and certainly not perfect, the TLVs were well on their way to becoming institutionalized and to making a contribution to occupational health practice. In 1961 TLVs were published in booklet form, and in 1962 the ACGIH published documentation of sources for the Threshold Limit Values. The TLVs provided guidelines at a time when few other guidelines or standards existed.

In the 1940s and 1950s, a database did not exist in toxicological inhalation science. As toxicologists began to carry out studies that would place standard setting on a more scientific basis, the Threshold Limits Committee utilized toxicological science to set TLVs.

The TLVs were certainly not perfect, and no one claimed them to be so. A volunteer committee of a professional association, composed of members who were either government employees (state, federal, or local) or employed by universities, and nonvoting consultants who usually represented private companies and occasionally organized labor recommended the TLVs.

In 1971, when the Occupational Safety and Health Act became effective, OSHA faced the task of issuing standards. The new agency had congressional authority to promulgate certain standards, referred to as *start-up standards,* without rulemaking. The start-up standards included already existing national consensus standards and established federal standards. OSHA issued start-up safety standards taken from standards issued by the American National Standards Institute (ANSI) and the National Fire Protection Association (NFPA). Congress considered both ANSI and NFPA to be national consensus organizations and their standards "consensus standards." OSHA also issued health standards in 1971. The health standards included the 1968 Threshold Limit Values (TLVs) of ACGIH for about 400 substances, already utilized under the Walsh-Healey Public Contracts Act. The TLVs, while not consensus standards, had been incorporated earlier into the Department of Labor Walsh-Healey Act and hence were federal standards and could be adopted by OSHA. The new OSHA standards were Permissible Exposure Limits (PELs), some based on inadequate documentation. Hence a private professional organization had an immense impact on public, legally mandated standards for occupational safety and health.

Although it had been stated many times in the introduction to the TLV booklet that the TLVs were not intended to be used as standards, ACGIH did not act to remind OSHA about the proper utilization of TLVs. The TLVs were never intended to be used as legal enforceable standards, a position that had always been taken by the Conference and the TLV Committee and presented in the TLV booklet since 1960.

ACGIH members seemed to have mixed emotions about use of the TLVs. They wanted to contribute to the new federal effort to bring about a healthy and safe workplace, and they were proud of the TLVs. Very little discussion can be found about this issue. One of the few such discussions, recorded at the 1972 ACGIH Business Meeting at the American Industrial Hygiene Annual Conference, gives some insight into the members' attitudes on the question of incorporating TLVs into federal OSHA law.

A member asked,

> Did anybody on the Board show any signs of annoyance or displeasure or concern over the fact that the TLVs are being used by the Labor Department in contravention to a resolution passed by this body at the 20th

Annual Meeting in 1958 at Atlantic City to the effect that the TLVs or any other such list should not be used in any code, rule, regulation or any such manner directly used in a regulation as a sole criterion for determination of a health hazard?

The answer:

Yes there was some concern expressed. . . . The impression I had was that the Board, regardless of whether or not they approved this course, felt that the limits would be used, and that if ACGIH TLVs were going to be used in this manner they should use what the ACGIH now considers to be their TLVs.[39]

A question was raised related to removing from the TLV booklet the section written to discourage incorporating TLVs into laws or codes.

This problem, it seems to me, is further complicated by the fact that the TLV list contains a discouragement from putting TLVs in the law, and that has been removed from the latest list, without any action on the part of the membership. Just on the part of the Committee itself, they took that action. I would like to know why they took it.[40]

The answer to that question was

There is nothing in my opinion, that ACGIH can do to prevent or stop anyone, any state or federal agency, from using our ACGIH TLVs in standards. This has been discussed for a number of years. There are arguments on both sides.

An answer from the floor was

I might like to say that at the time the TLV Committee did make the decision to drop or modify the statement that these shall not be used, it was done about a year ago, and was not a unilateral action by that Committee, by the Airborne Contaminants TLV Committee. It was brought up in discussion at the Board of Directors Meeting. Now, they may not have done correctly by not bringing it before membership as a valid point of concern, but I do want to point out that it was done with the knowledge of the Board of Directors.[41]

During the twenty years the Occupational Safety and Health Act has been in existence, OSHA promulgated only 24 full standards. Hearings for virtually every standard were lengthy, complex, and often vituperative.

Along with OSHA standards, TLVs continued to be utilized and criticized by representatives of industry, who believed the TLVs too stringent, and by labor representatives, who considered the TLVs too lax. The ACGIH continued to state its position that the TLVs were guidelines intended for use in the practice of industrial hygiene, to be interpreted and applied by industrial hygienists, and not to be utilized as legal standards.

Partly in response to the small number of OSHA standards passed from 1970 to 1988 and to the deregulation of the 1980s, OSHA began a project to update TLVs and to incorporate them as standards for PELs in the workplace. In this project, referred to as the PEL Project, OSHA intended to utilize unofficial guidelines (TLVs) as enforceable limits. Since 1970 the process by which OSHA permanent standards were promulgated depended on specific enabling legislation, under which the standard is promulgated by the appropriate administrative agency.

By law a permanent standard for a toxic chemical under the Occupational Safety and Health Act is also not merely a limiting concentration of the material in air, below which those dealing with the agent may be exposed. It has a broader interpretation and includes engineering controls, work practices, medical and environmental surveillance, and record keeping, as well as provisions for permissible exposure levels.

In light of the above, and in spite of the ACGIH position that TLVs were guidelines (not standards), intended for use in the practice of industrial hygiene, and to be interpreted and applied by industrial hygienists, in 1988 OSHA began to update the TLVs and incorporate them as OSHA standards. The ACGIH once again stated its position:

> These recommendations or guidelines (TLVs) are intended for use in the practice of industrial hygiene, to be interpreted and applied only by a person trained in this discipline. They are not developed for use as legal standards, and the American Conference of Governmental Industrial Hygienists (ACGIH) does not advocate their use as such.[42]

Although not exactly forceful, the statement did reiterate that the ACGIH never intended TLVs to be legal standards and did not agree with the philosophy of the OSHA PEL Project.

The PEL Project caused bitterness about the TLVs and the ACGIH to surface. The OSHA hearings on the PEL Project made that clear. Spokespeople from the labor movement angrily denounced the PEL Project, the TLVs, and the ACGIH.[43] The policy implication of the PEL project—a change in the rulemaking procedure—was unacceptable to the labor community and to some members of ACGIH. Labor took the position that OSHA allowed the ACGIH to usurp the role of the National

Institute of Occupational Safety and Health (NIOSH) in the rulemaking process.

> They [TLVs] are devised by a small group of individuals working with little technical support, based on limited literature reviews by a summary, informal, off the record process without clear decision rules or external appeal. Defenders of the process often say that this is the best that can be done under the circumstances, and that standards of this type are a necessary benchmark when no other process is available. Indeed if TLVs are viewed in that light, TLVs have served a purpose in the past and may continue to do so. However, the TLV list is often paraded as something more authoritative and balanced. This has been misleading and detrimental as well.[44]

Shortly before the OSHA PEL hearings, an article was published in the *American Journal of Industrial Medicine* that severely criticized the TLV Committee, the TLV Committee procedures, and its product, the TLVs. The authors criticized the TLVs for scientific inadequacies and for what they perceived as industry's role in the TLV process. They believed that corporate interests unduly influenced TLVs because of the use of corporate consultants, undocumented communications, and unavailable scientific documents. They discredited the TLV Committee by citing conflict of interest and concealment of corporate influence upon the TLV process, and they criticized the TLVs for outdated or inadequate documentation.[45]

In 1989 OSHA held hearings on the PEL Project, and the agency accepted approximately 400 PELs as standards. Some of the PELs have been questioned in the courts. In spite of ACGIH misgivings, the PELs remain adopted OSHA standards.

In retrospect, because of the role TLVs played in setting occupational health guidelines, controversy existed from the beginning. For almost 50 years a small group of professionals applied their expertise to solving occupational health problems by creating industrial exposure guidelines that were widely circulated and integrated into the practice of occupational health. In effect, a private organization set and implemented public policy for occupational health because their recommended guidelines were both respected and adopted. Although not perfect, TLVs made a major contribution to the practice of occupational hygiene and profoundly affected occupational health policy and practice. Yet controversy always existed, depending on whether guidelines were considered too high or too low. In either case, and according to what their interests were, either labor or management took issue. Both questioned how the TLVs were developed. They asked if a truly impartial group of scientists developed the TLVs without undue influence, for example, from industry. The questions of how the TLVs were to be utilized, and by whom, also caused controversy.

In spite of dissension, TLVs did make a major contribution to the practice of occupational hygiene and protecting workers. A volunteer committee with limited resources provided guidelines when no others existed. But when OSHA proposed to use TLVs in its PEL Project, it proposed substituting the voluntary efforts of a professional organization of experts for those of a legally mandated standard-setting agency.

The PEL Project changed and enlarged the impact of professionals by adopting updated TLVs and making them federally enforceable standards. OSHA set a precedent when it chose to rely on the work of a volunteer, technical specialty committee (the TLV Committee) rather than on OSHA's own experts and OSHA's mandate to set standards. The PEL Project made it impossible to view TLVs as removed from occupational health policy deliberations or outside the context of interest group politics.

SETTING MANDATORY
LEGAL OCCUPATIONAL
EXPOSURE STANDARDS

In 1970 Congress enacted sweeping occupational safety and health legislation. The revolutionary Occupational Safety and Health Act requires safe and healthful working conditions for approximately 65 million employees in an estimated 5 million establishments. The act represents a legal shift to direct, effective statutory prescriptions and administrative regulations. It recognizes that the health of workers is a public rather than private problem and that the standards to be observed are explicit public policy.

The Occupational Safety and Health Act created OSHA, a regulatory agency empowered to set and enforce regulations to control safety and health hazards. It is part of the Department of Labor and headed by an Assistant Secretary of Labor for Occupational Safety and Health appointed by the President. OSHA was to set mandatory safety and health standards, inspect workplaces to ensure compliance, propose penalties for violators, and provide for education for the public, workers, and employers. The Act also created NIOSH as a research agency and part of the Centers for Disease Control of the Public Health Service in the Department of Health and Human Services. The director of NIOSH would be appointed by the Secretary of Health and Human Services. Congress mandated NIOSH to conduct research and related activities to develop criteria or recommendations for OSHA to utilize for setting standards and to provide professional education and dissemination of health and safety information.

The heart of OSHA's activities is promulgating health and safety standards. These legally enforceable standards should be viewed as instruments of public policy. Once promulgated, OSHA standards must be adhered to, or civil and criminal penalties can be levied by the agency. Initially, an OSHA compliance officer could enter a workplace at any time and without warning to the employer. This procedure was challenged, and the Supreme Court ruled in *Barlow vs. OSHA* to restrict the right of entry of an inspector and to require a search warrant if the employer requested one.[46]

The Occupational Safety and Health Act departed from traditional landmark labor legislation (Fair Labor Standards Act, Taft-Hartley Act, Landrum-Griffin Act, and the Civil Rights Act of 1964). They spelled out a list of do's and don'ts for employers and unions to follow. The Occupational Safety and Health Act, rather than setting forth ground rule prohibitions, established the machinery to grind out the rules. The law empowered an administrative agency to issue detailed safety and health regulations, called *standards,* that would have the force and effect of law.[47] The format and procedures for adopting occupational safety and health standards were prescribed by Congress in the act. Their extent and specificity were not invented by the bureaucracy.

According to Section 6(a), the "Secretary shall promulgate the standard which assures the greatest protection of the safety or health of the affected employees."[48] The General Duty Clause, Section 5(a)(1), states that "each employer shall furnish to each of his employees employment and a place of employment which are free from recognized hazards that are causing or likely to cause death or serious physical harm to employees." Section 6(b)(5) states that "the Secretary shall set the standard which most adequately assures, to the extent feasible, on the basis of the best available evidence, that no employee will suffer material impairment of health or functional capacity even if such employee has regular exposure to the hazard dealt with by such standard for the period of his working life."[49] A great deal of debate since then has focused on the word *feasible.*

Congress also defined a standard in Section 6(b)(7).

Any standard promulgated under this subsection shall prescribe the use of labels or other appropriate forms of warning as are necessary to insure that employees are apprised of all hazards to which they are exposed, relevant symptoms and appropriate emergency treatment, and proper conditions and precautions of safe use or exposure. Where appropriate, such standard shall also prescribe suitable protective equipment and control or technological procedures to be used in connection with such hazards and shall provide for monitoring or measuring employee exposure at such locations and intervals, and in such manner as may be necessary for the protection of employees. In addition, where appropriate, any such standards shall prescribe the type and frequency of medical examinations or other tests which shall be made available, by the employer, or at his cost, to employees exposed to such hazards in order to most effectively determine whether the health of such employees is adversely affected by such exposure.[50]

The procedures for adopting standards are prescribed under the Administrative Procedures Act.

1. Notice of Intended Rulemaking: The agency indicates in the *Federal Register* that it is seeking information on a given subject because it intends to adopt a standard. Technically, one is disobeying the law if information in possession is not forwarded to the agency.
2. Informational Hearing: The agency can hold an informational hearing to review the "state of the art."
3. Issuance of a Proposal: The agency publishes in the *Federal Register* a proposal for the standard. In this proposal the agency has utilized the best information at hand to structure the best standard it deems possible. At the request of members of the public, a hearing on the proposal can be and usually is called. At the hearing those wishing to discuss specific ingredients of the standard do so and are open to question by members of the audience. An administrative law judge presides over the hearing, and a hearing record is maintained. After the hearing, 60 days are usually permitted for those wishing to enter comments into the hearing record. The hearing record can easily exceed 10,000 pages; in the case of the OSHA Generic Carcinogen Standard it exceeded 100,000 pages.
4. Promulgation of Standard: The agency attempts to promulgate a standard within 60 days after the posthearing commentary period. However, with the lengthy hearings to date, it has been very difficult for OSHA to meet this time frame. The courts have ruled that, depending on other priorities, the agency need not adhere to this time frame.
5. Effective Date of the Standard: The effective date of the standard is usually 60 to 90 days following promulgation of the standard in the *Federal Register*.

In addition to these steps, the agency can form an advisory committee early in its efforts to develop a standard. The activities of an advisory committee, such as those used early in the 1970s, add approximately 9 to 12 months to the time required to develop a standard. Without the formation of an advisory committee, OSHA could theoretically develop a standard in 27 months. However, the agency has not met this time frame in the past. Table 2-2 indicates the number of standards passed in the 20-year period from 1970 to 1990.

The statute permits OSHA to issue an emergency temporary standard in those instances where, as stated in Section 6(b)(8)(c), it is determined "that employees are exposed to grave danger from exposure to substances or agents determined to be toxic or physically harmful or from new hazards, and that such emergency standard is necessary to protect employees from such danger." Emergency temporary standards are in effect for 6 months. During that period the statute requires the agency to initiate proceedings for promulgation of a permanent standard. The history of emergency temporary standards has been spotty. The courts have placed great weight on the term *grave danger,* because the emergency temporary standard procedure bypasses due process. Our courts do not often approve of such procedures.

During the first 2 years that the statute was in effect, OSHA was permitted to adopt previously existing occupational safety and health standards that were present in federal law. Also, OSHA could adopt standards promulgated by consensus organizations.

The process of setting occupational safety and health standards was and still is value laden, highly controversial, and rife with conflicting principles

TABLE 2-2.   Summary of Permanent Health Standards Promulgated Since 1970

Standards Completed	Standards Proposed but Not Completed
Asbestos	Beryllium
Vinyl chloride	Sulfur dioxide
Arsenic	Ketones
Benzene	Toluene
Coke oven emissions	Ammonia
14 Carcinogens	MOCA 4,4'-methylene-2,2'-dichloroaniline
Lead	Trichloroethylene
Cotton dust	Generic monitoring
DBCP 1,2-dibromo-3-chloropropane	Generic medical surveillance
Carcinogen policy	
Acrylonitrile	
Ethylene oxide	
Asbestos	
Hazard communication ("right to know")	
Permissible Exposure Limits (revised)	
Access to medical and exposure records	
Noise abatement, hearing conservation	
Formaldehyde	
Field sanitation	

and processes. The issues associated with TLVs and OSHA standards clearly illustrate this. Each OSHA standard was subjected to both political and scientific scrutiny and took an inordinate amount of time to be promulgated. Examples of the issues and conflicts associated with passing a full OSHA standard can be found in chapters 7 and 8 of this book. Under the OSHAct promulgation of standards is slow. There are new demands to speed up the process, improve efficiency, and simplify the process. Perhaps a new chapter in the history of promulgating occupational safety and health standards will soon be written because of the growing consensus that the Occupational Safety and Health Act needs to be amended.

References
1. Corn, M. and J. Corn. "Setting Standards for the Public: An Historical Perspective." In *Impact of Energy Production on Human Health,* edited by E. C. Anderson and E. M. Sullivan. Proceedings of the LASL Third Life Sciences Symposium, Los Alamos, New Mexico, October 15–17, 1975. CONF-751022, National Technical Information Service. Washington: U.S. Department of Commerce, 1975.
2. Ibid.
3. Corn, J. *Environment and Health in Nineteenth Century America.* New York: Peter Lang, 1989.
4. Selikoff, Irving, Jacob Churg, and E. Cuyler Hammond. "The Occurrence of Asbestosis Among Insulation Workers in the United States." *Annals of the New York Academy of Sciences* 132(12): 139–155.
5. U.S. Department of the Interior. "Asbestos." *Bureau of Mines Mineral Yearbook* 1982.
6. National Cancer Institute. *Asbestos: An Information Resource.* DHEW Publication 79-1681. Washington: Department of Health, Education, and Welfare, 1978.
7. Dupre, J. Stefan. Chairman. *Report of the Royal Commission on Matters of Health and Safety Arising from the Use of Asbestos in Ontario* vol. 1. Toronto: Ministry of the Attorney General, 1984.
8. Ibid., 100–101.
9. Ibid., 98–100.
10. Auribault, M. "Note sur l'hygiene et la securite des ouvriers dans les filatures et tissages d'amiante." *Bulletin de l'Inspection du Travail* 14 (1906): 126.
10A. Murray H. Montague. "Statement Before the Committee in the Minutes of Evidence." In *Report of the Departmental Committee on Compensation for Industrial Disease* London: His Majesty's Stationery Office, 1907.
11. Cooke, W. E. "Pulmonary Asbestosis." *British Medical Journal* 2(1927): 1024–25.
12. Legge, T. *Industrial Maladies* London: Oxford University Press, 1934.

13. Lynch, K. M. and W. A. Smith, "Asbestos Bodies in Sputum and Lung." *Journal of the American Medical Association* 95(8): 659–661.

13A. Donnelley, J. "Pulmonary Asbestos." *American Journal of Public Health* 20 December 1933, 1275–1281.

13B. Ellman, P. "Pulmonary Asbestosis: Its Clinical, Radiological, and Pathological Features and Associated Risk of Tuberculosis Infection." *Journal of Industrial Hygiene* 15 (July, 1933): 165–183.

14. Lanza, A. J. ed. *Silicosis and Asbestosis.* New York: Oxford University Press, 1938.

14A. Lanza, A. J. and J. A. Goldberg, *Industrial Hygiene.* New York: Oxford University Press, 1939.

14B. Clark, W. I. and P. Drinker, *Industrial Medicine.* New York: National Medical Book Company, 1935.

14C. Legge, T. *Industrial Maladies.* London: Oxford University Press, 1934.

14D. Rosenau, Milton J. *Preventive Medicine and Hygiene,* 6th ed. New York: Appleton-Century-Crofts, 1935.

15. Lynch, K. M. and W. A. Smith, "Pulmonary Asbestosis III. Carcinoma of Lung in Asbesto-Silicosis." *American Journal of Cancer* 14(1935): 56–64.

16. Doll, Richard. "Mortality from Lung Cancer in Asbestos Workers." *British Journal of Industrial Medicine* 12(1955): 81–86.

17. "Biological Effects of Asbestos," *Annals of the New York Academy of Science* 132. (December 1965).

18. Teleky, L. *History of Factory and Mine Hygiene* New York: Columbia University Press, 1948.

19. Ibid., 130.

20. Ibid., 131.

21. Bowditch, M., C. K. Drinker, P. Drinker, H. H. Haggard, and A. Hamilton. "Code for Safe Concentrations of Certain Common Substances in Industry." *Journal of Industrial Hygiene and Occupational Medicine* 22(June 1940): 251.

22. Dreessen, W. C., J. M. Dallavalle, T. I. Edwards, J. W. Miller, and R. R. Sayers, *A Study of Asbestosis in the Asbestos Textile Industry.* Public Health Bulletin 241, U.S. Public Health Service. Washington: U.S. Government Printing Office, 1938.

23. Ibid., ix.

24. Ibid., 91.

25. American Conference of Governmental Industrial Hygienists, *Documentation of TLVs.* Cincinnati: ACGIH, 1962.

26. Ibid., 17–19.

27. Corn, J. K. and M. Corn, "The History of Accomplishments of the Occupational Safety and Health Administration in Reducing Cancer Risks." In *Reducing the Carcinogenic Risk in Industry,* edited by P. F. Deisler New York: Marcel Dekker, 1984.

28. Department of Labor, Occupational Safety and Health Administration, *Federal Register* 51(119): 22614.

29. Ibid., 22615.
30. Ibid., 22612.
31. Letter from H. Dyktor to C. Pool, 9 May 1940 (sent to the Executive Committee of NCGIH). Archival materials ACGIH, Cincinnati.
32. Transactions of the Eighth Annual Meeting of the American Conference of Governmental Industrial Hygienists, Chicago, 7–13 April 1946.
33. Transactions of the Tenth Annual Meeting of the American Conference of Governmental Industrial Hygienists, Boston, 27–30 March 1948.
34. Ibid.
35. Ibid.
36. Transactions of the Fifteenth Annual Meeting of the American Conference of Governmental Industrial Hygienists, Los Angeles, 8–11 April 1953.
37. Transactions of the Sixteenth Annual Meeting of the American Conference of Governmental Industrial Hygienists, Chicago, 24–27 April 1954.
38. Transactions of the Twentieth Annual Meeting of the American Conference of Governmental Industrial Hygienists, Atlantic City, NJ, 19–22 April 1958.
39. Transactions of the Thirty-fourth Annual Meeting of the American Conference of Governmental Industrial Hygienists, San Francisco, 14–19 May 1972.
40. Ibid.
41. Ibid., 19.
42. Letter from Arvin Apol, Chair, ACGIH, to Docket Officer, United States Department of Labor. Docket Number H202.
43. Testimony on OSHA's PEL Update Proposal by Michael Silverstein, M.D., M.P.H., Assistant Director, Health and Safety Department, International Union, UAW; Franklin E. Mirer, Ph.D., C.I.H., Director, UAW, Health and Safety Department; Rafael Moure-Eraso, Ph.D., C.I.H., Industrial Hygienist, Health and Safety and Environment, IUD, AFL-CIO, 5 August 1989.
44. Testimony of Franklin Mirer on OSHA's PEL Update Proposal, 5 August 1988.
45. Castleman, Barry I. and Grace E. Ziem, "Corporate Influence on Threshold Limit Values." *American Journal of Industrial Medicine* 13(1988): 531–559.
46. Mintz, Benjamin. *OSHA: History, Law and Policy.* Washington: Bureau of National Affairs, 1984.
47. *The Job Safety and Health Act of 1970.* (Washington: Bureau of National Affairs) 1971.
48. Ibid., 96.
49. Ibid., 97.
50. Ibid., 98.

Chapter 3

Risk Assessment and Federal Policy, 1970–1990

By the 1970s—and coincident with the advent of the Occupational Safety and Health Act—the uncertainties and complexities in identifying and assessing occupational disease, rapid proliferation of new hazards, better ability to quantitate, the increased role of regulation, and interest generated by new laws, the press, public interest groups, and the labor community placed occupational health in the public eye and concentrated interest on policy decisions and how they would be arrived at. Controversy often focused on the selection of priorities for those hazards that would be controlled and the degree of control.

The federal government's expanded regulation of health, safety, and environment stimulated a vigorous national debate about risk. The debate began with United States policy for regulating carcinogens in the environment and eventually led to a shift in occupational and environmental health policy regarding acceptance by government agencies of risk assessment as an analytical tool for regulatory decision making.

During the past 20 years definitions of risk have been developed in an endeavor to create a framework for regulatory decision making. Risk assessment is noteworthy as a key concept (albeit one hardly mentioned just 20 years ago) now utilized by decision makers in the United States. Risk has been explored extensively in the social science literature. Themes such as scientific uncertainty, mixing and confusion of science and politics in decision-making efforts, controversy within the scientific community and the larger community, and the problems associated with the political use of science have all been explored and described.[1, 1A–1E]

Contemporary concern with risk focuses on health risks caused by modern industrial processes rather than on natural phenomena such as earthquakes and hurricanes. The movement toward more clearly defining health risks began in the 1970s when federal regulatory agencies started shifting toward the utilization and acceptance of risk assessment as an analytical tool for regulatory decision making. It signaled one of the most far-reaching changes in United States regulatory policy. By the late 1980s most of the occupational health community had come to terms with the concepts of occupational health risk assessment.

Today the suggested national approach to priority selection is risk assessment. Because decisions are made within both social and scientific frameworks, addressing risk involves two extremely different kinds of activities: assessing risk, an empirical, scientific activity, and judging risk, a normative, political activity. The first implies objectivity, whereas the second implies social value.[2]

Increasing pressure was placed on regulators to justify proposed regulations. What benefits would be gained? What risks would be removed or reduced? What would it cost? Numerous problems and limited resources made difficult decisions even more difficult. Decision making for regulating occupational hazards is still evolving, but it is now accepted that control and prevention of occupational hazards is, in part, an exercise in sorting out the risks associated with a particular agent.

This chapter addresses the following questions: What is risk assessment? What is the relation between risk assessment and government decision making for occupational health? How and why did risk assessment become the primary instrument for regulatory decision making in the United States?

THE IDEA OF RISK

Human beings have always had to deal with *risk,* defined here as potential for harm, or the unwanted negative consequence of an event, or the probability of something harmful happening. It is an old idea and one that has been redefined and refined over the years. In an article published in 1985, Covello and Mumpower reviewed the history of "risk analysis" and "risk management." The authors said they intended to provide a basis for future directions in risk by presenting a historical perspective; they also indicated the development of the idea of risk assessment.[3] They emphasized historical development prior to the twentieth century and presented information on the intellectual antecedents of risk. For example, they discussed the role of soothsayers who served as consultants for risky, uncertain, or difficult decisions in the ancient world, the development in the seventeenth century of probability theory (calculus) as a tool to analyze quantitatively, theories of random frequencies, and the need for record keeping before statistics or probabilities could be discussed and analyzed. They made the point that modern risk analysis had twin roots in the mathematical theories of probability and the scientific methods for identifying causal links between adverse health effects and hazardous activities. Covello and Mumpower also reviewed risk management techniques or strategies for managing risk. They included insurance, common law, direct government intervention (epidemics, food contamination, building and fire codes, etc.), and voluntary self-regulation. They discussed changes between past and present. Table 3-1 indicates nine changes from past to present pointed out by Covello and Mumpower.

TABLE 3-1.   Changes from Past to Present Relevant to Risk Analysis

Change	Example
Shift in the nature of risk	From infections to chronic disease
	Growth in number of auto accidents
Increase in life expectancy	Female born in U.S. in 1900: life expectancy, 51; female born in U.S. in 1975: life expectancy, 75
	Male born in U.S. in 1900: life expectancy, 48; male born in U.S. in 1975; life expectancy, 66
Increase in new risks	Chemicals
	Radioactive wastes
	Pesticides
	Nuclear accidents
Increase in scientists' ability to measure and identify risks	Advances in laboratory testing
	Epidemiological methods
	Computer simulations
Increase in the number of scientists and analysts whose work is focused on health, safety, and environmental risks	Risk assessment and analysis is emerging as an identifiable discipline
Increase in the number of formal quantitative risk analyses produced and used	Use of highly technical quantitative tools
Increase in role of federal government in assessing and managing risks	Increase in number of health, safety, and environmental laws
	Increase in number of federal agencies charged with managing health, safety, and environmental risks
Increase in participation of special interest groups	Risk assessment has become increasingly politicized
Increase in public interest, concern, and demands for protection	

RISK ASSESSMENT DEFINED IN THE 1990s

Risk assessment is a regulatory strategy. It is a response to the combination of scientific uncertainties and conflicting political and economic interests seemingly inherent in decision making to control environmental hazards to human health. It is the "characterization of the potential adverse health effects of human exposures to environmental hazards."[4] Risk assessment, usually synonymous with quantitative risk assessment (QRA), includes: (1) description of potential adverse health effects based on the evaluation of results of epidemiological, toxicological, clinical, and environmental research; (2) extrapolation from those results to predict the type and estimate the extent of health effects in humans under given conditions of exposure; (3) judgments as to the number and characteristics of persons exposed at various intensities and durations; (4) summary judgments on the existence and overall magnitude of the public health problem; and (5) characterization of the uncertainties inherent in the process of inferring risk.[5] Broadly defined, risk assessment is a method or process of calculating the estimated likelihood that a particular exposure to a substance will cause illness such as cancer.[6] Risk assessment is a tool developed to approach priority setting. It is an evolving science that involves value judgments.

Risk assessments start from the common concept that the probability of harm from exposure to a toxic substance is a function of two variables: the exposure or potency and the dose of the substance. When assessing the magnitude of a risk, risk assessors need a method to relate the dose-response curve (information about incidence of disease produced by different doses of a substance) and the exposures that humans incur or are likely to incur. Risk assessors utilize a number of different mathematical models to relate dose-response to exposure.

Criticism of risk assessment includes the often wide disparity in risks predicted by different models, the lack of empirical verification of the models, and the uncertainties inherent in extrapolating from animals to humans or from high dose to low dose.

Supporters of risk assessment acknowledge the difficulties noted by critics but defend risk assessment as a tool developed to approach priority setting. They view risk assessment as a way to structure decisions that inevitably involve uncertainties. They believe that efforts to reduce human exposure to toxic substances require a method to estimate the magnitude of health consequences of alternative control methods.

The contemporary concept of risk can be further refined and subdivided into risk management and risk communication. Journals and professional organizations devoted to the field of risk analysis include the Society of Risk Analysis and *Risk Analysis*. The subject has been studied and written about mainly in the scientific and political science literature. It has become what we might refer to as a "hot topic" and indicates the growth of the field. *Risk management* is the process of developing policies to manage risk based on the risk assessment. Risk management is extremely complex because it includes legal, political, ethical, social, economic, and other value considerations. Regulatory agencies perform risk management under a variety of legislative mandates to develop, analyze, and compare regulatory options and to select appropriate regulatory responses to a potential health hazard.[7] Risk management involves value judgments on issues such as acceptability of risk and reasonableness of the cost of control.

Risk communication is the exchange of information about risk among scientists, decision makers, and the public. The three *R*s—risk assessment, risk management, and risk communication—are often referred to as risk analysis and can be viewed as a response to controversial, complex occupational and environmental health issues. The three *R*s represent the need to develop prudent public health policies to regulate risks that are not negotiable by individuals and therefore require governmental intervention.

REGULATION

What we currently refer to as *occupational risk assessment* has roots in the early twentieth century, beginning with the recognition that occupational disease was associated with exposures to hazards in the workplace, for example, lead, mercury, and silica. At the time, the general approach to risk assessment was characterized by acceptance of the premise that health risk was related to the degree of exposure and the toxicity of the chemical. As risk assessment evolved, it included epidemiological data of worker populations for a specific hazard and collected dose-response data that involved animal tests.

TOXICOLOGICAL SCIENCES

Much of the data utilized for risk assessments depends on the toxicological sciences. The development of one branch of toxicology, inhalation toxicology, is briefly described here to illustrate one aspect of risk assessment. Today, inhalation toxicology is a major factor in assessing both environmental and occupational exposures. Key questions asked to determine occupational and environmental exposure limits are, What concentrations pose significant hazards to health? What concentrations can be accepted without undue risk to health? The importance of available technical information in these determinations cannot be overestimated. The quality of risk assessment and therefore regulatory decisions is largely dependent on the adequacy of available information.

Traditionally, *toxicology* has been defined as the science or knowledge of poisons. *Inhalation toxicology* is the science or knowledge of inhaled poisons or toxicants. Its subject areas include (1) the physical and chemical characteristics of material in the air, (2) the basic biology of the respiratory tract, (3) the deposition and retention of inhaled materials in the body and their interaction with critical biological units, and (4) how such interactions with the respiratory tract and other systems produce disease. Answers to these questions provide a basis for assessing the health risks of airborne materials. Concern for the effects of inhaled toxicants can be traced back for centuries. For example, smoke and odors from coal aroused attention in London in the thirteenth century, and early in the Industrial Revolution the steam engines and the utilization of coal to fuel them also caused concern. Despite early problems, air pollution and industrial poisonings were seldom thought of as serious issues. The field of inhalation toxicology as we know it today can be traced back only decades.

By the mid-twentieth century, air pollution incidents began to create public awareness that airborne materials could produce disease. Three well-known smog episodes occurred, the first in 1948 in Donora, Pennsylvania; the second in 1952 in London; and the third in 1962 in London. All three episodes resulted in marked increases in morbidity and mortality from respiratory effects. Before the smog episodes, except for some early pioneering occupational studies by Theodore Hatch, Cecil Drinker, and Philip Drinker, there was meager interest in the occupational or environmental health hazards of airborne materials, and very little research carried out in the field of inhalation toxicology. In spite of toxicology's long history, not until research was carried out in the twentieth century did our knowledge of how air pollution and industrial pollution produced disease begin to increase. After World War II, chemical agents, radioactive materials, and automotive exhaust emissions stimulated concern for health effects caused by airborne materials.

Interest in health effects of radioactive materials grew during World War II because of the Manhattan Project (the United States effort to develop the atomic bomb). Stoddard Warren directed medical services at the Manhattan Project, where concern arose about the toxicity of uranium. Warren turned to colleagues at the University of Rochester, Stockinger and Hodge. The result was the University of Rochester Atomic Energy Project, where pioneering work on inhalation toxicology of uranium and other materials used in the Manhattan Project resulted.

World War II also ushered in emphasis on quantitation. With the use of radioactive materials and their precise measurements, it became possible to quantitate the amount of radioactive material in the air and the fraction deposited and retained in the body and to relate the observed biological effects to various measures of dose.

The increased awareness that air pollution could be caused by chemicals and motor vehicles also heightened interest in inhalation toxicology and in the need to understand health effects of airborne materials in occupational and environmental exposures, and thus the need to limit the potential for induction of disease.

The history of the technology of the early years of inhalation toxicology describes how hard-won some of the advances were. Early experiments from the nineteenth century through the mid-twentieth century illustrate the long road to advances in the techniques associated with inhalation risks.

Inhalation toxicology is devoted to determination of the toxic effects on animal species following carefully controlled inhalation of toxic material(s). Thus it depends on the rigor with which the investigator standardizes the exposed animal population and controls for the delivery of the toxicant over what are usually extended periods of time under controlled conditions of temperature and pressure.

Some of the early developments in the hardware and techniques of inhalation toxicology include advances in inhalation chamber technology, aerosol and gas generation systems, and exposure characterization. Advances in measurements of effects in the exposed animal populations, although equally significant, are not discussed here. Only one aspect of inhalation toxicology, chamber design and toxicant delivery systems, is discussed in this chapter.

The experimental study of the effects of airborne agents (gases, vapors, aerosols) can be traced back to the nineteenth century, when investigators began to develop exposure techniques and equipment to meet their needs for researching the effects of airborne toxicants on animals. One such early recorded experiment occurred in 1865. Eulenberg described controlled animal inhalation exposure studies. He used a cubical wooden chamber, 12.75 inches on a side, with two glass walls for the exposure of small laboratory animals to high concentrations of numerous toxic and asphyxiant gases. The inner walls of the chamber were coated with varnish, which contained rubber. Auxiliary equipment included a gasometer for measuring flow rates and a manometer for measuring chamber pressure. The airflow was driven by water placement.[8]

In 1875–76 Van Jns reported the case of a carefully designed chamber for dust exposure of small laboratory animals. The chamber was a

20 cm × 20 cm × 10 cm total enclosure. The dust feed was an ingenious arrangement of a mechanical shaker on a dust-filled funnel, from which the dust was dispersed by a motor-driven bellows into the exposure chamber. The exposure chamber was used to study the effects of the inhalation of diatomaceous earth.[9]

Lehman and his associates published a number of papers on inhalation toxicity based on work done at the Hygienic Institute in Munich. Inhalation chambers were used for dynamic exposure of cats, rabbits, guinea pigs, and frogs. Reports on inhalation experiments by the group described an inhalation chamber for gases (1911, Dubitzki), dust exposure unit (1912, Saito), a mist exposure unit (1912, Lehman, Saito, and Majima), and additional gas and vapor units (1913, Lehman and Hasegawa). They studied the effects of arsenic hydride, white lead, and nitrogen oxides.[10] These early investigators made a definite distinction between exposures of animals to particulates and exposures to gases because the chambers required for dust exposures posed the more difficult problems of dust generation, distribution, and control.

Other investigators used chambers for dust exposure. In 1918 Mavrogodato used a simple wooden box to expose guinea pigs to coal, shale, flue, and flint dust. A two-bladed electric fan dispersed the material contained in a wooden trough. Dust concentrations varying from 27,000 to 45,000 mg per cubic meter were determined by inserting a cotton-plugged tube in the side of the chamber and withdrawing a volume of dusty air.[11] A chamber utilized by Gardner (1920) was a box containing animals on trays in the upper portion and in the bottom a barrel of finely divided granite that was agitated by a paddle wheel. To compensate for the variation of dust concentration with location, the animals were placed in different positions each day.[12]

In 1929 Sayers and associates designed a 250-cubic-foot chamber used at the Pittsburgh Experiment Station of the United States Bureau of Mines to expose guinea pigs to static concentrations of halogenated hydrocarbons. Concentrations were established by pouring the desired amount of liquid onto a large, flat surface in the chamber. Distribution was aided by a fan, and air samples were taken at regular intervals throughout the exposure.[13] Similar experimental studies continued through the 1930s by investigators such as Yant, Schrenk, Sayers, and Gross.

A 1932 handbook on animal experimentation methods in occupational medicine reviewed techniques of animal exposure. Areas discussed included inhalation of dusts, fumes, droplets and fogs, vapors, and gases, as well as feeding, skin absorption, and injection.[14] Experiments were carried out at the DuPont Haskell Laboratory, United States Bureau of Mines, Division of Industrial Hygiene of the United States Public Health Service, and the Dow Chemical Company.

In 1940 Fairhall and Sayers described a chamber used by the Division of Industrial Hygiene, USPHS. This box-type chamber had heavy glass fronts fitted against soft rubber gaskets, which could be removed to allow for cleaning and transferring animals. The dust feed passed through an allu-triator to provide a uniform dust dispersion. Air samples of the chamber atmosphere were taken at a rate of 1 cubic foot per minute (cfm) through filter paper disks mounted on a side wall. The samples could be analyzed chemically and microscopically for concentration particle size and composition.[15]

Apparatus and methods for testing toxicity continued to develop. By the 1940s organic solvents as well as dust were being studied and principles influencing design and operation of constant-flow chambers for gas and vapor inhalation exposures were discussed. The data presented allowed for formulas derived for predicting equilibration time for chamber concentration, effects of airflow, chamber size, the character and quantity of the interior surface, the shape of the chamber, the relative areas of air inlet and door opening, and the number and size of animals on chamber concentrations equilibrium times, surface effects, and animal loadings.

By mid-century there was a first generation of exposure units for inhalation toxicology studies. The task remained to improve these chambers with regard to distribution of toxicants to the animals, uniform rate of toxicant delivery (particularly dusts), and improved characterization of the toxicant (that is, the description of the exposure). These developments continued and were necessitated by the investment in the 1950s, 60s, and 70s in major inhalation facilities by the Defense Department (Wright Patterson Air Force Base), the Department of Energy (Lovelace Foundation and Batelle Northwest), the United States PHS-EPA Human Inhalation Facilities at Chapel Hill and Ranchos Los Amigos Hospital in Los Angeles, the NIH (NIEHS), and selected private institutions such as the Chemical Industry Institute of Toxicology and DuPont's Haskell Laboratories.

The major investment of government funds to support increased regulatory effort in the 1960s and 1970s rapidly advanced the state of the art. Today the state of the art permits an investigator to expose animal or human subjects to well-characterized, consistent challenge atmospheres of toxicant gas or particles. Mixed exposures (multiple agents) are still rare and probably represent the next stage of development. Complex mixtures are rarely generated for inhalation toxicology. When interested in complex mixtures, toxicologists dilute the "real" thing, such as diesel exhaust or smog. Studies still focus on single or at most two toxicants in air.

The scientific foundation for risk assessment illustrated here for one area, inhalation toxicology, has been developing for most of the twentieth century.

RISK ASSESSMENT AFTER
WORLD WAR II

Soon after World War II, toxicological sciences emerged as a tool to assess chemical risks. The extraordinary growth of the chemical industry had led to rapid proliferation of new hazards and the need to identify occupationally and environmentally caused disease. In 1962 the publication of Rachel Carson's *Silent Spring* focused popular and scientific interest on environmental pollution caused by the widespread use of chemicals.[16] The Food and Drug Administration (FDA) also began to study and identify chemicals that could be safely added to food and drugs.

A number of federal environmental laws were passed in the late 1960s and the 1970s. These laws dealt with toxins as well as carcinogens. Many laws were passed in response to the belief that the majority of cancers are environmental and occupational in origin and that most of them could be prevented by reducing human exposure to carcinogenic substances. Rushefsky attempted to explain and explore values inherent in making societal decisions by addressing the evolution of federal cancer policy and the role of risk assessment in regulatory action.[17] In a short period the federal government put in place a large number of detailed regulations and policies to control occupational and environmental health risks.

Table 3-2 lists major environmental and occupational legislation. It encompasses an enormous area of environmental concerns including air, water, pesticides, consumer protection, and occupational safety and health. Note that most of the laws were passed between 1969 and 1980.

The earlier environmental laws were concerned with air, water, and pesticides. The later laws were more oriented toward human health. In the 1970s, with the shift in environmental policy to regulate carcinogens and toxic substances, risk assessment began to gain acceptance as an analytical tool for regulatory decision making.

The numerous regulations of the 1970s treated each risk differently and allowed for little consensus on a risk framework and on the associated degree of protection. For example, the legislation employed three different risk frameworks.

1. Risk-based laws provided for regulations to reduce risks to zero without considering other factors.
2. Balancing laws implied some degree of risk above zero and the balancing of risks against benefits.
3. Technology-based laws directed regulatory agencies to impose specific levels of control.[18]

TABLE 3-2.   Major Legislation, 1938–1986

Acronym	Act	Date
FDCA	Federal Drug and Cosmetics Act	1938
FIFRA	Federal Insecticide, Fungicide, and Rodenticide Act	1948, 1972, 1975
FHSA	Federal Hazardous Substances Act	1966
NEPA	National Environmental Protection Agency Act	1969
PPPA	Poisonous Packaging Prevention Act	1970
OSHA	Occupational Safety and Health Act	1970
CAA	Clean Air Act	1970, 1977
FWPCA	Federal Water Pollution Control Act (now Clean Water)	1972, 1977
MPRSA	Marine Protection, Research and Sanctuaries Act	1972
CPSA	Consumer Product Safety Act	1972
FEPCA	Federal Environmental Pollution Control Act	1972
SDWA	Safe Drinking Water Act	1974, 1977
HMTA	Hazardous Materials Transportation Act	1974
RCRA	Resource Conservation and Recovery Act	1976, 1979
TSCA	Toxic Substance Control Act	1976
SMCRA	Surface Mine Control and Reclamation Act	1977
UMTCA	Uranium Mill Tailings Control Act	1978
CERCLA	Comprehensive Environmental Response, Compensation and Liability Act	1980
SARA	Superfund Amendments and Reauthorization Act	1986
AHERA	Asbestos Hazard Emergency Response Act	1986

It is not surprising that policy makers were confronted with a balancing process. Guidelines for testing and evaluating experimental data and for choosing substances to be regulated caused controversies that have been documented in the literature on policy. Scientific uncertainties and lack of the information needed to form a basis for regulatory action created controversy over the magnitude and management of health risks. In fact, environmental legislation opened a Pandora's box full of conflicts, divergent views and the inability to separate scientific from political inputs during the decision-making process. Furthermore, both proponents and opponents of regulation showed they were capable of distorting science to achieve their objectives.[19, 19A] Both references illustrate the relationship between science and policy decisions and the possibility of distortion of science to achieve objectives.

Many problems associated with occupational health and environmental policy decisions could be attributed to the capabilities and limitations of science. The scientific basis for risk decisions has always been a pervasive, persistent issue in regulatory legislation. That, combined with the controversy over goals and how to achieve them, became central factors in shaping policy. At the same time the environmental issues that emerged in the 1970s fast became health issues.

The number of health oriented laws increased in the 1970s, and emphasis shifted from environmental protection issues to environmental health risks. The focus of health risk in occupation and environment changed from diseases of short duration (infectious diseases) to long-term, life-threatening effects of disease, for example, cancer. Lower concentrations of chemicals caused concern, based on the concept that low levels of dose over long periods of time could cause disease, and on the concept of latency. Sophisticated instrumentation and analytical techniques made it possible to measure lower exposures. All of these factors, coupled with the growth of public interest groups and industrial lobbies and a changing political and economic climate that prompted balancing costs against risks, contributed to the change in direction. Some regulators believed that a new emphasis on a more analytical and reasoned basis to determine the magnitude and nature of health risks associated with chemicals might moderate the conflicts and political controversy over goals and the means to achieve them.

Of course, many factors can account for the evolution of federal risk policy. Some have already been presented. For the remainder of this chapter, three events that signaled a change in outlook and profoundly affected United States environmental policy are discussed.

1. Publication of William Lowrance's influential *Of Acceptable Risk*[20]
2. The Supreme Court decision that invalidated the Occupational Safety and Health Administration's Benzene Standard
3. Publication of the National Academy of Sciences' *Risk Assessment in the Federal Government: Managing the Process*[21]

OF ACCEPTABLE RISK

In 1976, understanding that a major policy challenge faced regulators and decision makers who struggled with the problem of how to evaluate risk, William Lowrance defined risk for regulators, scientists, decision makers, and the public. He pointed out the central role of risk assessment as a tool for organizing and analyzing information and indicated its scientific basis and limitations. Lowrance posed the following questions:

How did we determine how hazardous these things are? Why is it that cyclamates one day dominate the market as the principal calorie-cutting sweetener in millions of cans of diet drinks, only to be banned the next day because there is a "very slight chance" they may cause cancer? Why is it that one group of eminent experts says that medical x-rays (or food preservatives, or contraceptive pills) are safe and ought to be used more widely, while another group of authorities, equally reputable, urges that exposure to the same things should be restricted because they are unsafe? At what point do debates such as that over DDT stop being scientific and objective and start being political and subjective? How can anyone gauge the public's willingness to accept risks? Why must there be these endless controversies over such things as lead, whose effects on health have been known in detail for years? Are people being irresponsible, or is there something about these problems that just naturally spawns confusion? Just what sort of a decision making tool is this notion of "safety"?[22]

He then explored the problems, the underlying concept of safety itself, and the general features of the social context within which safety decisions are made. He also emphasized the contributions of scientists and technically trained people.

Lowrance differentiated between the evaluation or estimate of risk, which is a scientific endeavor, and the acceptance of levels of risk, which involves sociopolitical decision making. Addressing risk therefore involves two extremely different kinds of activity. First, measuring risk involves measuring the probability and severity of harm. It is an empirical, scientific activity and implies objectivity. Second, judging risk involves judging the acceptability of risk. It is a normative, political activity and implies value. Lowrance wrote, "Safety is not measured. Only when those risks are weighed on the balance of social values can safety be judged: a thing is safe if its attendant risks are judged to be acceptable."[23]

The mixing of the two activities—the assessment of risk and the judgment of risk—in United States occupational health standard setting since 1971 had caused much bitterness among identifiable groups associated with standard setting. One need only study the history of occupational safety and health standard setting since the Occupational Safety and Health Act was passed. No standard was passed without conflict. Often because there were insufficient scientific data to assess risk, final recommendations reflected judgment of the acceptability of risk, not a combination of measurement and judgment.

THE BENZENE DECISION

In February 1978 the Occupational Safety and Health Administration issued a permanent standard for benzene that lowered the previous permissible exposure limit adopted in 1971 from 10 parts per million (ppm) to 1 ppm, with a ceiling of 15 during any 15-minute period.[24] OSHA adopted the stringent 1 ppm based on its "lowest feasible level" policy. There were no animal data on benzene, and epidemiological data indicated that human beings contracted leukemia at concentrations significantly higher than the prior level of 10 ppm. OSHA concluded that it was not possible to demonstrate a threshold level for benzene-induced carcinogenicity or to establish a safe level for benzene exposure. They therefore decided that the permissible exposure level to benzene should be reduced to the lowest feasible level. OSHA concluded that higher exposures carry greater risk; because they could not make a determination of what constitutes a safe level of exposure to benzene that presents no hazard, they could not answer the question of whether a safe level of exposure to benzene exists.[25] OSHA said that prudent health policy required they limit exposure to the maximum extent feasible.

The permanent standard on benzene was challenged in court by the American Petroleum Institute (API) and vacated. In 1980 the United States Supreme Court upheld the lower court's decision to vacate the standard. Although the court was divided, the plurality concluded that OSHA must show that the toxic substance at issue (benzene) created a "significant" health risk in the workplace and that the lower standard would eliminate or reduce that risk. They said that OSHA had failed to make a valid determination that reducing the permissible exposure level of benzene from 10 ppm to 1 ppm is reasonably necessary to protect workers from a risk of leukemia.[26]

Several aspects of this decision need to be emphasized. First, when the plurality struck down OSHA's policy of regulating carcinogens to the lowest feasible level, they said the policy was based on assumptions rather than evidence. Second, the plurality held that an agency was required to show by substantial evidence that "at least more likely than not" the new standard would eliminate or reduce a significant risk in the workplace. Third, the plurality said that the risk from a substance must be quantified sufficiently to enable the Secretary to characterize it as significant in an understandable way.

The bottom line is that, in the era after the Benzene Decision, the test of significant risk and utilization of quantitative risk assessment are now part of the process of setting standards to regulate toxic chemicals. The Supreme Court's Benzene Decision profoundly affected future regulatory policy.

RISK ASSESSMENT IN THE FEDERAL GOVERNMENT: MANAGING THE PROCESS

The third event that affected the development of risk assessment was the publication in 1983 of the National Academy of Sciences report *Risk As-*

sessment in the Federal Government: Managing the Process. The report suggested that risk assessment and the choice of regulatory options be distinguished from each other. The committee that wrote the report recommended

> The regulatory agencies take steps to establish and maintain a clear conceptual distinction between assessment of risks and consideration of risk management alternatives; that is, the scientific findings and policy judgments embodied in risk assessment should be explicitly distinguished from the political, economic, and technical considerations that influence the design and choice of regulatory strategies.

They also recommended the development of uniform inference guidelines for the use of federal agencies in the risk assessment process and that risk assessment methods be established for regulatory agencies.[27]

Other factors that led to incorporation of risk assessment in regulatory decisions include

1. New methods for risk assessment began to appear in the scientific literature. There was significant growth in this area of scientific inquiry in the 1970s.[28]
2. Publication of an Interagency Regulatory Liaison Group document on the subject of risk assessment, 1979.
3. Publication of the Science and Technology Policy document on the subject of risk assessment, 1985.
4. Publication of the EPA document, *Risk Assessment and Management: Framework for Decision Making (1984).* EPA Administrator William D. Ruckelshaus embraced the idea of risk assessment.
5. Executive Order 12044 (1978) required regulatory agencies to perform an analysis for regulations that would have significant economic impacts.
6. Executive Order 12291 (1981) replaced 12044 and required all executive agencies to prepare a regulatory impact analysis to help select regulatory approaches in which potential benefits to society from regulation would outweigh potential costs. Economic impact statements were required, as was detailed regulatory analysis. These federal guidelines were aimed at "regulatory reform".

LOOKING BACK

When the United States embarked on its ambitious road of regulating environmental and occupational risks to reduce them, it quickly became clear that these activities were intimately associated with subjective aspects of the political process and that a method was needed to select priorities. The Benzene Decision, for example, reflected the need for risk assessment as a method for selecting priorities. The scientific community and the policy community responded with the set of techniques they called *risk assessment*.

Although enormous progress has been made in risk assessment during the past 20 years, many problems remain unsolved, especially how to create a bridge of understanding between risk assessment and political decisions concerning types, levels, distribution, and acceptability of risk. The challenge remains of how to make intelligent decisions about health risks. It was hoped that the introduction of risk assessment into the decision-making process would lead to priority setting and informed decisions based on the best scientific estimates of risk.

References
1. Zimmerman, Rae. *Governmental Management of Chemical Risk*. Lewis Publishers, 1990.
1A. Nelkin, Dorothy. *The Language of Risk*. Beverly Hills: Sage Publications, 1985.
1B. National Research Council. *Improving Risk Communication*. Washington: National Academy Press, 1987.
1C. Jasanoff, Sheila. *Risk Management and Political Culture*. New York: Russell Sage Foundation, 1986.
1D. Paustenbach, Dennis J. *The Risk Assessment of Environmental Hazards*. New York: John Wiley & Sons, 1987.
1E. Lave, Lester B., ed. *Risk Assessment and Management*. New York: Plenum Press, 1987.
2. Lowrance, William. *Of Acceptable Risk*. Los Altos, CA: William Kaufman, 1976.
3. Covello, Vincent T. and Jeryl Mumpower. "Risk Analysis and Risk Management: An Historical Perspective." *Risk Analysis* 5(2): 103–120.
4. National Academy of Sciences. *Risk Assessment in the Federal Government: Managing the Process*. Washington: National Academy Press, 1983.
5. Ibid., 18.
6. Mintz, Benjamin W. *OSHA History, Law and Policy*. Washington: Bureau of National Affairs, 1984.
7. National Academy of Sciences. *Risk Assessment*.
8. *Exposure Chambers for Research in Animal Inhalation*, Public Health Monograph 57. Washington: U.S. Department of Health, Education and Welfare, 1959.
9. Ibid.
10. Ibid.
11. Mavrogodato, A. "Experiments on the Effects of Dust Inhalation." *Journal of Hygiene* 17(1918): 439–459.

12. Gardner, L. U. "Studies on the Relation of Mineral Dusts to Tuberculosis." *American Review of Tuberculosis* 4(1920): 734–755.
13. Sayers, R. R., W. P. Yant, B. G. H. Thomas, and L. B. Berger, *Physiological Response Attending Exposure to Vapors of Methyl Bromide, Methyl Chloride, Ethyl Bromide and Ethyl Chloride,* Public Health Bulletin 185 Washington, U.S. Public Health Service, 1929.
14. *Exposure Chambers,* 45.
15. Fairhall, L. T. and R. R. Sayers, *The Relative Toxicity of Lead and Some of Its Common Compounds,* Public Health Bulletin 253. Washington, U.S. Public Health 1940.
16. Carson, Rachel. *Silent Spring.* Boston: Houghton Mifflin, 1962.
17. Rushefsky, Mark E. *Making Cancer Policy.* Albany: State University of New York Press, 1986.
18. Office of Technology Assessment. *The Assessment of Technologies for Determining Cancer Risks from the Environment.*Washington: Congress of the United States, Office of Technology Assessment, 1981.
19. Corn, Jacqueline K. "Vinyl Chloride, Setting a Workplace Standard: An Historical Perspective on Assessing Risk." *Journal of Public Health Policy* 5 (December 1984): 497–512.
19A. Bayer, Ronald, ed. *The Health and Safety of Workers: Case Studies in the Politics of Professional Responsibility.* New York: Oxford University Press, 1988. Both illustrate the relationship between science and policy decisions and the possibility of distortion of science to achieve objectives.
20. Lowrance. *Of Acceptable Risk.*
21. National Academy of Sciences. *Risk Assessment.*
22. Lowrance. *Of Acceptable Risk,* 2.
23. Ibid., 8.
24. Federal Register 43. 5918 (1978). *OSHA Occupational Exposure to Benzene: Permanent Standard.*
25. Federal Register 43. 5931, 5946 (1978).
26. IUD v. API, 448 U.S. 607 (1980).
27. National Academy of Sciences. *Risk Assessment.*
28. Paustenbach, Dennis J. *The Risk Assessment of Environment Hazards.* New York: John Wiley & Sons, 1989.

Case Studies

Based on two premises, first, that past events affect present policies and, second, when historical perspective is ignored change is viewed in purely personal terms, and judgment made on the basis of individuals not issues, the second part of this book consists of case studies that present historical perspective for five occupational hazards. The five hazards include lead, asbestos, vinyl chloride, cotton dust, and silica. The case studies offer insight into issues associated with making occupational health policy decisions and understanding factors that fostered change in the past.

Efforts to control occupational disease depend on social and historical factors as well as scientific and technical factors. Themes in the case studies include the relation and interaction between science and politics, the social determinants of occupational health, the relation between priority setting and allocation of resources, benefit versus risk in decision making for hazardous materials, and the utilization of science and technology. Each case study explores one or more of these themes.

Chapter 4

Lead

Concepts of health risk associated with lead usage have undergone profound change. In the past, observations of the relationship between lead usage and lead poisoning were severely limited by lack of knowledge and the relatively small quantity of the metal used. The benefits of lead usage, when contrasted with the risk incurred, outweighed the known hazards.

Industrial growth contributed to the growing interest in lead poisoning when a significant increase in the number of workers at risk generated new interest in an old disease. A concept of control developed based upon the principle of a dose-response relationship. The old idea of benefit versus risk remained because technological society needed lead but agreed that by controlling the factory environment lead could be used safely. The criteria for injury remained clinical plumbism.

Growth and change in lead utilization, once again, brought lead hazard to the forefront. New applications distributed lead throughout the environment exposing both workers and the general population. A segment of the scientific community now suggests that the old criteria to judge risk, clinical symptoms of plumbism, are inadequate.

Lead, essential to modern technology, has long been a critical raw material. In the past the benefits of lead usage, when contrasted to the associated risks, outweighed the known hazards. The growth of technology and its by-product, an urban industrial society, created an ever-increasing demand for lead and resulted in an inordinate number of occupational exposures to lead in industry. The dangers to health associated with continued exposure to certain forms of lead cover an entire spectrum of symptoms, with the most advanced form referred to as *plumbism*. In the twentieth century the efforts of workers in the field of industrial health, a changed social climate, and new legislation made the lead industries relatively safe for workers. The injuries to health caused by lead poisoning were considered technologically controllable, and the cost of control was accepted to ensure the continued availability of an essential industrial raw material.

At early stages of scientific and technological growth that involve the use of needed materials, we often accept the chance of jeopardy to health incurred by utilization of hazardous materials. In many cases, the perils are unknown. As technology becomes more refined, we become aware of more subtle hazards; and, as social philosophy changes, our attitudes toward risks also become more humane. This new awareness of danger and the concomitant change in social philosophy may cause a reevaluation of the continued utilization of the material in terms of benefit versus risk.

Because of the growth of science and technology, the concept of health risk associated with lead usage has changed. It is no longer confined to severe, acute, or chronic lead poisoning of industrial workers. Our sophisticated technological society has evolved a new set of criteria to judge risk; they include contamination of the total environment. A corollary of scientific and technological progress, the ability to make more accurate and sensitive measurements, has focused attention on small differences in trace quantities of lead in biological systems and in our environment.

The purpose of this chapter is to provide historical perspective for the current controversy related to the definition of lead poisoning by showing the relationship between increased lead usage, increased medical understanding of lead hazard and the disease of plumbism, increasing ability to make more accurate measurements, and changing social attitudes toward the disease of lead poisoning itself. By examining the changing attitudes toward the assumption of risk in the use of lead, I hope that the dialogue between the two schools of thought on health effects of lead usage will be clarified. Because lead is but one of a number of toxic materials used in industry, its story may give some insight into other problems associated with needed but toxic materials used in industry and reveal the complexity of evaluating the use of such materials. Lead is an example of materials utilized by modern industrial society that possess highly toxic properties. Other materials on which attention has been focused recently are asbestos and mercury.

PROPERTIES AND USES OF LEAD

Lead (Pb), a blue-gray metallic element with a high degree of plasticity, is widely distributed throughout the world. The principal lead ore is galena (PbS); other common ores are lead cerussite ($PbCO_3$) and anglesite

(PbSO$_4$). Since ancient times, lead has been used for making pipe, paint pigments, and ceramics. Industrial growth resulted in the utilization of lead in thousands of products in our society.

Because of its plasticity and softness, lead can be rolled into sheet and foil. It can be made into rods, pipes and tube containers. Lead is used in building construction for roofing, cornices, tank linings, electrical conduit, water pipes and sewer pipes. It is used for yacht keels, plumbbobs and sinkers in diving suits. Antimonial lead is a major type metal. Lead-antimony alloys are used for accumulator plates, cable coverings, ornamental castings, and the filling of bullets for small arms ammunition. Soft solder, used for soldering tin plate and lead pipes, is an alloy of lead. Lead base alloys are used in engineering to make bearing metals.[1]

Lead is the basic ingredient in the solder that binds together our electronic miracles and it is the sheath that protects our intercontinental communications systems. It is the barrier that confines dangerous x-rays and atomic radiation, i.e. "shielding material." It is sound proofing for buildings and ships and jet planes. It is the major component in the batteries that start our cars and it is the gasoline that runs our cars.[2]

HEALTH EFFECTS

New uses have been found for lead and old ones discarded, but the danger of disease stemming from lead's toxic properties has persisted. Moderns refer to the effects of exposure to lead as *plumbism*. In other eras it has been called *saturnism, colic, dry gripes, dry bellyache,* and *potter's palsy*. Until recently we knew only the acute or chronic manifestations of plumbism. In the discussion that follows, these ailments are referred to as the *clinical effect* of lead. In the decade beginning with 1960, more subtle manifestations of lead hazard became the subject of much speculation and concern within the medical-scientific community. The subtle manifestations of lead intoxication are referred to as the *subclinical effects* of lead on humans.

Lead can enter the body through the respiratory system or by way of the gastrointestinal tract. Certain lead compounds can also be absorbed through the skin. Lead in sufficient dosage can affect the gastrointestinal, excretory, nervous, and circulatory systems.[3] Acute clinical manifestations of plumbism are intestinal colic (the most common and most painful), lead encephalopathy, attacks of coma, delirium, and convulsions. Chronic forms of plumbism are mental dullness, inability to concentrate, poor memory, headache, deafness, wrist drop, a blue line on the gums, and transitory pains in muscles and joints.[4] Because the only requirement for possible incorporation of lead into the body is exposure to the metal or one of its compounds, lead poisoning can affect the general public as well as industrial workers.

Ancient people used lead without understanding its toxic properties. They mined and reduced lead from ores by smelting and parted silver, commonly associated with lead, by cupellation prior to 2000 B.C.[5] Leaden objects have been found in Egyptian tombs. The Assyrians used rods and lumps of lead for currency. In Greece and Rome, lead had a variety of uses: as a component of water pipes and solder, to extract silver from gold and copper, and to glaze pottery. The Assyrians, in about 700 B.C., used lead oxide as a base for the first glaze that would adhere to baked clay.[6] Yellow and black pigments were made with lead. White lead for white pigment was made at least as early as the fourth century B.C. by a process that lasted in a modified form until the twentieth century.[7]

Although in this early age of metal usage there did exist some association between lead and disease (Hippocrates in 370 B.C. described a severe attack of colic in a man who extracted metals and recognized lead as a cause of the symptoms), the ancients paid little heed to the diseases associated with lead usage. By modern standards the ancients used an infinitesimal amount of lead. When disease did occur, it remained largely ignored because workers were slaves and forced laborers.

During the Middle Ages lead was used for cisterns, pipes, roofing, extraction of silver and gold from copper, pottery glaze, and to make clear glass. The economic and technological development of this period is reflected in the fact that lead poisoning affected miners and metal workers. Hunter noted "the possibility that occupational factors could be of importance in explaining a given illness was ignored all through the dark ages."[8] Perhaps because plumbism affected only a few members of a submerged class, it received little attention.

The amount of lead produced and its utilization grew and changed with the centuries. By the sixteenth century Agricola and Paracelsus imparted some information related to the diseases of lead miners and lead workers. In 1556 Agricola published his classic encyclopedia treatise on the mining and metal industries, *De Re Metallica*. Although the work described mining and metallurgy processes, there are incidental but excellent descriptions of illness and accidents to workers, particularly in Book Six. Agricola wrote about mining, smelting, assaying, and extracting ores, including lead. The arts and sciences necessary for a miner to be acquainted with included, according to Agricola, "medicine, that he should be able to look after his diggers and workmen, that they do not meet with these diseases, to which they are more liable than workmen in other occupations, or if they do meet them, that he himself may be able to heal them or may see that doctors do so.'' Agricola also noted that mining is a "perilous occupation to pursue, because miners are sometimes killed by the pestilential air they breathe. . . . Of the illnesses, some affect the joints, others the lungs, some the eyes and finally some are fatal to men.''[9]

In 1567, 11 years after the publication of *De Re Metallica,* a book written by Theophrastus Bombast von Hohenheim, usually known as Paracelsus, was published entitled *Von der Bergsucht und anderen Bergkrankheiten* (On Miner's Sickness and Other Miner's Diseases). Paracelsus presented his observation of diseases of miners and the effects of various minerals and metals on the human organism. In this unique sixteenth-century work, Paracelsus recognized two groups of diseases that affected miners and smelterers: diseases of the respiratory organs and pathologic conditions resulting from ingesting or inhaling poisonous metals. He recognized poisonous effects of various metals and differentiated between acute and chronic poisoning. Paracelsus wrote,

> The miner's sickness is . . . the disease of the miners, the smelterers, the pitmen and others in the mines. Those who work at washing, in silver or gold ore, in salt ore, in alum and sulfur ore or in vitriol boiling, in lead, copper, mixed ores, iron or mercury ores, those who dig such ores succumb to lung sickness, to consumption of the body, and to stomach ulcers, these are known to be affected by miner's sickness.[10]

In 1614 Martin Pansa, a physician, wrote and published *Consilium Peripneumoniacum,* "a faithful guide in the troublesome mine and lung sickness, in which is presented the most important causes of such ailments, the poisonous ones, which arise from the mine, as well as the common ones which come from the fluxes, before that, however, how many may be compared with the small world and with the mine, and finally, how diseases are to be expelled."[11] The book proposed to teach miners how to protect themselves from the diseases to which they were subject. Pansa wrote, "All those who dig gold or silver ores, salt, alum, sulphur, lead, copper, tin, iron or mercury are subject to it [disease]."[12] He described causes of disease, symptoms, and treatment.

In the seventeenth century, Vernatti, a physician, noted the effects of lead poisoning on white lead makers. He observed, "The Accidents to Workingmen are Immediate paine in the stomach, with exceeding Contortions in the Guts and Costiveness that yields not to Cathartics."[13]

Bernadino Ramazzini, a physician and professor at the University of Modena and later Padua, made a significant contribution to occupational medicine in the eighteenth century. He considered occupational health socially important, undertook a study of the morbid conditions caused by certain occupations, and called attention to the practical application of his knowledge. He stressed the need to study the relationship between occupation and disease and to account for social and occupational factors, as well as medical factors, when dealing with patients. Ramazzini said,

> For we must admit that the workers in certain arts and crafts sometimes derive from them grave injuries, so that where they hoped for a sustenance that would prolong their lives and feed their families, they are too often repaid with the most dangerous diseases and finally, uttering curses on the profession to which they had devoted themselves, they desert their post among the living. [14]

Ramazzini wrote of 54 different occupations associated with lead poisoning and described some of them in the sections concerning potters and painters.

> During the process [i.e., potting] or again when they use tongs to daub the pots with molten lead before putting them into the furnace, their mouths, nostrils, and the whole body take in the lead poison that has been melted and dissolved in water; hence they are soon attacked by grievous maladies. First their hands become palsied, then they become paralytic, splenetic, lethargic, cachectic and toothless, so that one rarely sees a potter whose face is not cadaverous and the color of lead.

With regard to treatment of workers, Ramazzini noted, "It is hardly ever possible to give them any remedies that would completely restore their health. For they do not ask for a helping hand from the doctor until their feet and hands are totally crippled and their internal organs have become very hard; and they suffer yet from another drawback, I mean they are very poor." [15]

In the nineteenth century Tanquerel de Plances observed that the characteristic traits and primary effects of plumbism could be observed in men who worked in an atmosphere of lead and dust fumes.

At the same time that physicians associated lead with toxic effects on workers in mines and industries, it also affected unsuspecting members of the general population. "Dry gripes" was associated with one of colonial America's early industries, rum distillation. Distillation in lead worms and still heads contaminated large quantities of rum. In 1723 Massachusetts colony passed "An Act for Preventing Abuses in Distilling of Rum and Other Strong Liquors in Leaden Heads or Pipes," perhaps to ensure that New England would not lose its rum market. [16] Pewter, an alloy of tin and lead, was commonly used in the seventeenth and eighteenth centuries;

pewter utensils such as teapots, mugs, creamers, tankards, platters, basins, and pitchers came into contact with food. Pewter and lead poisoning were often characterized by the common ailment "dry gripes." Periodicals carried advertisements for cures. One such ad read,

> For the Good of the Public, a certain Person hath a secret Medicine which cures the Gravil and Cholick, immediately, and Dry Belly Ache in a little time; and restores the Use of the limbs again, (tho of never too long continuance) and is excellent for Gout. Enquire Mr. Samuel Gerrish, Bookseller, near the Brick Meeting House, over against the Town House in Boston, N.B. The Poor who are not able to pay for it may have it gratis.[17]

The Devonshire colic epidemic, one affecting the general public, raged for approximately 100 years until, in 1776, a physician, George Baker, published an essay entitled, *Concerning the Cause of the Endemial Colic of Devonshire*. The dangerous disease, colic, was common in Devonshire, a cider-drinking area of England. Baker's essay demonstrated that the "colic" was attributable to lead poisoning. Lead entered the bodies of the cider-drinking population of Devonshire because of extensive, reckless use of lead in the apparatus of cider making and storage. Baker determined that large millstones were hewn into segments and bound together with lead. At the end of his essay, Baker stated,

> May I presume to hope, that the present discovery of a poison, which has for many years exerted its virulent effects on the inhabitants of Devonshire, incorporated with their daily liquor, unobserved, and unsuspected, may be esteemed by those who have the power, and who have opportunities to remove the source of such much mischief, to be an object worthy of their most serious attention.[18]

Although it was known since ancient times that lead had toxic properties, physicians used a variety of lead preparations for therapeutic purposes. During the eighteenth and nineteenth centuries, physicians believed that lead in proper dosage could cure many ills. A medical dictionary published in 1745 discussed the medicinal virtues of lead.

> Both in its crude state and in all preparations, lead seems to be cooling, thickening, repelling, absorbing, and contracting so as to retard the circulation of the blood, hinder all secretions and hurt the nerves, by causing spasms, convulsions, trembling, difficulty in breathing and suffocation. Whence it appears unfit for internal use in any large dose; and, accordingly its medical uses are principally external.[19]

The author of the dictionary then suggested that lead, dissolved in a mild acid such as vinegar, might be used to cure running and ulcerous sores or skin diseases. Some encyclopedias of the eighteenth and nineteenth centuries counseled cautious use of medicinal lead. The *Encyclopaedia Britannica* stated, "The internal use of lead is certainly dangerous, though it is often prescribed in medicine; and even the external use of it is not altogether safe."[20] Rees' encyclopedia (1819) cautioned that the remedy often proved worse than the disease but suggested that externally applied ointments and plasters containing lead had a sedative, drying, and repellent quality.[21] A medical treatise written in the early nineteenth century claimed that lead therapy could cure consumption, diabetes, dysentery, and epilepsy. The author stated that "lead like all other powerful medicines when given in too large quantities, becomes a poison; but, we have the authority of many respectable physicians, for asserting that its cautious internal exhibition, may be practiced with perfect safety, and frequently with the greatest advantage to the patient."[22] In the mid-nineteenth century lead was still being used to treat consumption.[23] It is a paradox that while the hazards of this metal, both to the working population and the general public, grew increasingly apparent, it still retained its place as a medicine.

All of the observations of the association between lead usage and the risk of poisoning were severely limited by a lack of knowledge of the disease proper and the still relatively limited use of lead.

Lead usage reflects the growth of technology and industry. As industry grew, so did a "demand for a wide range of metals to meet new needs and the growing ability of industry to provide them."[24] By the first quarter of the twentieth century, more than 110 industries used lead in some form.[25] When the hazard associated with lead affected large numbers of people, it received more attention. Nineteenth-century problems of a growing industrial and urban society were so vast and the industrial workers were beset with so many problems (such as horrible factory conditions, low wages, long hours, child labor, and economic insecurity) that lead poisoning was not foremost among their grievances.

RENEWED INTEREST IN PLUMBISM IN THE TWENTIETH CENTURY

Four factors account for renewed and vigorous interest in health hazards due to lead in the twentieth century: increase in lead usage, more data relative to the incidence of lead poisoning, better medical understanding of lead hazards and the disease proper, and a change in social attitudes. In the twentieth century pioneers in the field of industrial health, well aware of the great incidence of plumbism in American mines and factories, spear-headed a battle for health legislation to provide more salubrious factory conditions. They made valuable contributions to the knowledge of the action of lead on the body, the prevalence of occupational plumbism in various trades, the period of exposure before poisoning occurs, the different forms the intoxication assumes, and the dangers that must be guarded against to protect workers in lead trades. In their classic work, *Lead Poisoning and Lead Absorption,* Thomas Legge and Kenneth Goadby used experimental and statistical evidence to show that inhalation of dust caused lead poisoning. In his preface to the volume, the editor stressed the extensive utilization of lead by industry.

> This volume deals with a subject of wide interest, for lead is dealt with in so many important processes of manufacture—in the making of white lead; pottery glazing; glass polishing; handling of printing type; litho-making; house, coach and motor painting; manufacture of paints and color; file making; tinning of metals; harness making; manufacture of accumulators, etc. . . . Lead intoxication, on account of the great variety of industries in which the metal or its salts are used, is the most prevalent of occupational diseases.[26]

During the first two decades of the twentieth century, a few deeply interested individuals stimulated government to take some action to control lead poisoning in industry. Alice Hamilton, a physician and one of the most outstanding reformers, pursued medical problems and also stressed the human and social problems associated with health conditions in industry. She described herself as a pioneer in an unexplored field of American medicine, the field of occupational diseases. In 1910 she started her investigations of lead hazards as a member of an Occupational Disease Commission appointed by Governor Deneen of Illinois. In 1912 she undertook a similar survey for the United States government. Alice Hamilton studied the following lead-associated industries: manufacture of white lead, litharge, orange mineral, and red lead; glazing and decorating of white ware and sanitary earthenware and tiles, i.e., art and utility ware, tiles, and porcelain; lead mining; lead smelting and refining; the manufacture of storage batteries and accumulators; the painter's trade; and the manufacture of rubber goods. In her autobiography she wrote of her concern with plumbism: "Every article I wrote in those days, every speech I made, is full of pleading for recognition of lead poisoning as a real serious medical problem."[27]

Efforts of pioneers in the field of industrial medicine, social legislation, the rise of unions, scientific knowledge, and new engineering devices have all tended to reduce the hazard of lead. Successful efforts to "clean up" lead industries have made the evils of excessive industrial lead poisoning an event of the past.

The successful efforts employ two essential principles of in-plant safety. The first is that there is a systematic dose-response relationship between severity of exposure to the hazard of lead poisoning and the degree of response in the exposed individual. As the level of exposure goes down, there is a graded decrease in the risk of injury. The risk becomes negligible when exposure falls below certain tolerable levels. These principles assert that agents such as lead can be dealt with safely and therefore do not have to be eliminated completely to keep industrial workers safe:

> The dramatic successes over the past half century in the control of occupational diseases in many industrial establishments attest to the validity of these principles [stated above] and the very operation of many chemical and other plants in which there is, of necessity, some contact with potentially dangerous agents is justified only on the grounds of continuing demonstration that such work situations are not, in fact, impairing the worker's health.[28]

This, then, is the philosophical basis for ideas that prevail today in industrial health. The hazards of lead poisoning are considered controllable to an acceptable degree because lead is essential to industry. In the opinion of the majority of the professionals in occupational health, the risks, because of control methods, have been reduced to a tolerable level.

CHANGING CONCEPTS OF LEAD HAZARD

However, in the 1960s, scientific, social, and economic changes and a proliferation of lead products resulted in a new concept of risk associated with the usage of lead in our society. This concept raises questions relative to subclinical effects of lead in humans. A segment of the professional community is suggesting that the criteria used to judge the risk, that is, clinical symptoms of plumbism, are inadequate. The arguments and data introduced by the advocates of this new philosophical and scientific approach to lead usage in our society will now be examined.

The modern concept about lead hazard stems from new and varied uses of lead that distribute it as never before throughout the environment and expose the worker and the general population. Trace quantities of lead can be found in food, beverages, and the air we breathe. The sources are industry, smelting, incineration, combustion of coal, and motor vehicles. These trace quantities of lead permeate all aspects of the environment, including vegetation and small living animals. Motor vehicles became an increasing source of lead in the atmosphere after tetraethyl lead was introduced as an additive to gasoline in 1923. Figure 4-1 indicates annual United States consumption of lead from 1926 to 1963.[29] The changes in the uses of lead later in this period are indicated in Figure 4-2.[30] Major lead usage shifted from metal products and paints to storage batteries and gasoline additives.

In the past, perhaps the most widely accepted criterion for lead intoxication was the concentration of inorganic lead in the blood, usually expressed as "mg lead/100 g whole blood." The basis for this criterion is that lead is stored in the body, primarily in the calcareous portion of the bone, and the blood lead is in equilibrium with the lead storage depots. Hence, as increased quantities of lead are stored in the body, the blood lead level increases. The physician has been guided by various levels of blood lead as indicators of pathophysiological changes associated with lead intoxication.

As an example, a concentration of 0.08 mg lead/100 g of whole blood has been the guideline for concern in industrial workers. It should be noted that clinical symptoms are very rarely associated with this level of lead. Although the sensitivity of the population at risk differs, clinical symptoms do not usually appear until a level of 0.15 mg lead/100 g whole blood is reached.[31]

The advocates of the modern risk concept of lead suggest that certain long-term health effects may occur in the general population exposed to lead from the many sources in the community cited previously. They ask the following questions: What is the blood-lead level of the general population? For example, in 1965 the Public Health Service published a study that is often referred to, *Survey of Lead in the Atmosphere of Three Urban Communities,* which measured blood lead levels in three cities. Concentrations of lead in blood seldom exceeded 0.05 mg per 100 g of whole blood.[31A] The concern of the advocates of the new philosophy of risk has nothing to do with the old criterion of blood lead, and one might say that an entirely new concept of lead injury has been introduced.

Because of the proliferation of lead products and the resulting introduction of lead into all aspects of our environment, the ability to make more accurate and sensitive measurements, and a more concerned attitude toward the welfare and health of the general population, new questions relative to subclinical effects of lead in humans have been raised. These effects are not measurable by any of the guidelines previously used. This new concept of risk, which focuses on trace amounts of lead in vegetation and food supplies, questions the long-term effects of lead in our ecological system. Research performed by Chow and Johnstone, Claire Patterson, and Harriet Hardy illustrate this new concept.

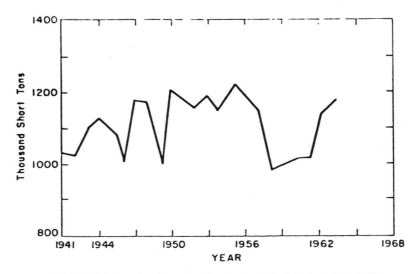

FIGURE 4-1. Lead consumption in the United States, 1926–1963.

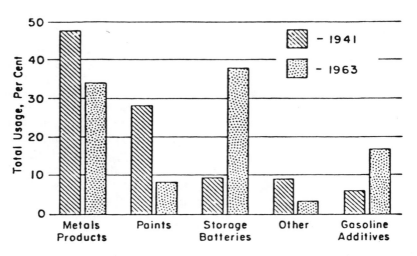

FIGURE 4-2. Shifts in the major uses of lead, 1941 versus 1963.

80

Chow and Johnstone determined the isotopic composition of lead in antiknock gasolines and in aerosols in the air of the Los Angeles basin. They also compared the lead-isotope compositions of the aerosols and rural snow from the Lassen Volcanic National Park in California. The isotopic composition of lead in 1963 rural snow was similar to that of Los Angeles basin aerosols. Therefore, they concluded that the contamination of the snow was attributable to industrial pollutants. "The high concentration of lead in water of precipitation and that of the surface of the sea can be attributed to automobile exhaust."[32]

Claire Patterson studied contaminated and natural lead environments and concluded that residents of the United States were undergoing severe chronic lead insult, that the geochemical relationships and material balance considerations show high ingestion of lead that is excessive compared to "natural" levels, and that the lead probably originates from lead dispersed in forms of lead alkyls, lead arsenates, food-can solder, and millions of tons of lead accumulated and stored in past paints, alloys, piping, glazes, and spent ammunition. Patterson, a geochemist, raised the question of blood lead levels. He stated that "existing blood lead concentrations have for decades been regarded as natural, although it is well known that the average value lies only slightly below the threshold levels for classical lead poisoning."[33]

Patterson then suggested (1) that natural and toxic blood lead levels be defined with greater care than in the past and (2) that deleterious effects of severe chronic lead insult be further investigated. The dialogue centers on levels of lead in blood and urine and on whether or not long-term exposure to lead leads to chronic ailment.

An epidemiological study of blood levels in the general population and in occupationally exposed groups suggested that increases in atmospheric lead will result in predictably higher blood levels in exposed populations. The study further stated,

It is clear for many urban residents that the total quantity of lead absorbed from the respiratory tract is of the same order of magnitude as that absorbed from the gastro-intestinal tract, that increased respiratory exposure within the range observed in community air pollution is capable of producing materially increased storage of lead in the body as reflected in the blood lead level, and that further increases in atmospheric level will result in higher blood lead levels in the population in a predictable relationship.[34]

Harriet Hardy, a physician, stated that one approach to examining the potential hazards from low-level exposures to lead was to study identifiable groups in the population, for example, infants. Because growing tissue is more vulnerable to lead than adult tissue, smaller doses are required to produce diagnosable lead poisoning in children. She also noted that there is a likelihood that lead in the body at levels considered to be harmless from industrial experience produces damage, and that if this is true, then other insults natural or man-made must be assessed in judging what amount of lead is harmful.[35] The influence of lead on cell growth, the effects of lead poisoning on mental development, the effects of lead on life span, and the concentration of lead in bone are some of the other questions raised.

The public has been alerted to the dangers of lead ingestion by young children who often eat chips of lead paint that fall from ceilings and walls of substandard and old buildings in ghetto areas. Prior to 1940, lead-based paint was commonly used for interior painting, and many old and deteriorated houses have thick layers of lead paint.[36] In children the minimum blood concentration at which poisoning occurs is lower than in adults. In 1971 the United States Public Health Service recommended that a blood lead concentration of 40 mg or more per 100 ml of whole blood is evidence suggestive of undue absorption of lead.[37] In 1969 it was estimated that the number of children in the United States with high blood levels of lead is possibly as high as 225,000.[38] In the case of childhood lead poisoning, it is believed that the problem is already defined and its causes known. Therefore, the public health attitude toward this problem is favorable. For example, the United States Public Health Service was actively involved in developing a program to control lead poisoning in children. In this case one might expect less controversy than in questions raised by atmospheric pollution. But in the 1990s childhood lead poisoning is fast becoming a major environmental issue.

At the same time that this new concept of lead injury was being explored and redefined, a number of other researchers continued to study lead concentrations in blood and reached the conclusion that there was not a significant health threat in the 1960s or in the foreseeable future. In a presentation at the thirty-first annual meeting of the Industrial Hygiene Foundation, Gordon Stopps and colleagues concluded that "there has not been an upward trend over the past 30 years in either urine or blood lead levels despite a very large increase in the use of gasoline containing antiknock agents."[39] An international study of normal levels of lead in blood and urine by the World Health Organization and the Columbia University School of Public Health and Administrative Medicine tested 30 subjects in each of 15 collaborating laboratories throughout the world and suggested normal levels of lead in blood and urine. These levels were compared with tabulated levels over the period 1925 to 1965. The results stated, "Of particular interest is the fact that there is no significant change in the normal blood and urinary level over the past four decades."[40]

As a result of these discussions, lead in the environment became a controversial public issue. In January 1965 a study of atmospheric lead in three selected cities was published. Samples of air were taken in locations in each city, and amounts of airborne lead were determined by laboratory analysis. Average concentrations of lead in all samples were 1.4 mg per cubic meter of air in Cincinnati, 1.6 in Philadelphia, and 2.5 in Los Angeles.[41] Concentrations of lead in blood of some 2,300 individuals in the three cities were determined. Eleven persons showed concentrations of 0.06 mg per 100 grams of blood, that is, six parts in 10 million. Concentrations of 0.08 mg of lead per 100 grams of blood are associated with toxic lead poisoning.[42] The 1964 United States Public Health Service study stated,

Even though the lead levels for the entire group, with a few exceptions, are within the presently accepted normal range in humans, the data show some interesting gradations. Levels of lead in the blood tend to increase gradually as the place of residence varies from rural to central urban areas. A second gradation is related to increasing opportunity for occupational exposure to exhaust of automobiles. . . . There appears to be an orderly progression in values [blood levels] according to the most likely concentrations of lead in the atmosphere to which these groups were exposed. For example, values are lowest for the suburban and rural groups, intermediate for downtown employees, and highest among those working with motor vehicles, such as drivers of cars and parking attendants. Although some groups, e.g., notably garage mechanics, are exposed to lead occupationally, they are tested to show comparative levels of lead in the blood. The likelihood that the variability of the amounts of lead ingested with food and beverage accounts for some of the differences between the lowest and highest of the groups, while relatively large, are well within the presently accepted range of lead levels for humans and are not significant in terms of a threat of the occurrence of lead intoxication.[43]

In December 1965, the United States Public Health Service sponsored a symposium on environmental lead contamination that brought together a wide range of views. Two major points of view were expressed. One group apparently felt that there is nothing to be alarmed about until the levels of lead in the blood of the ordinary population have reached the levels at which acute and clinically recognizable lead poisoning occurs. Others felt that any rise in the level of lead in the blood is hazardous.[44]

In June 1966, hearings were held before a Senate subcommittee on air and water pollution, with much discussion of lead as a factor in air pollution. The testimonies given summarized the problem and the positions of the opposing groups. Senator Edmund Muskie, chairman of the subcommittee, presided at the hearings. Robert A. Kehoe, a physician, professor emeritus of occupational medicine at the University of Cincinnati, and former director of the Kettering Laboratory, appeared as the most articulate spokesman of the group that believed there was no cause for alarm until concentrations of lead in the blood of members of the general population exceed levels at which clinically recognizable lead poisoning occurs. In his opening statement to the subcommittee, Kehoe said,

> The fact is, however, that no other hygienic problem in the field of air pollution has been investigated so intensely, over such a prolonged period of time, and with such definitive results. . . . Nevertheless, it is clear that this specific set of problems has been brought to such a point of understanding, in relation to public health, so as to remove it from the realm of urgency and to consign it into the group of hygienic problems on which a watchful and effective surveillance should be kept.[45]

Kehoe cited the results of the three-city survey to indicate that no increase in lead levels in blood occurred and said that his own research since 1939 was evidence that there was no increase in lead intake.

Felix E. Wormser, a consultant, former president of the Lead Industries Association, and retired vice president of St. Joseph Lead Company, testified that lead in air did not affect the general population and cited the same evidence referred to by Kehoe.

Other authorities viewed the problem of lead differently. Claire Patterson noted the discrepancy between what he termed *inferred natural values* of lead and *typical values*. He stated that scientists need more information on the biochemical mechanism of traces of lead in the body:

> No one can show, because they don't have the knowledge, what does happen. We don't describe the biochemistry involved. All we can do is describe the morphology and from microscopic slides you can look at the pictures of the destroyed tissue and this is about all. I would say that classical lead poisoning represents one extreme of a continuum of reaction of an organism in the human to various levels of exposure from lead.[46]

Harriet Hardy, assistant medical director of the Occupational Medical Service, Massachusetts Institute of Technology, was asked by Senator Muskie, "Is it accurate to describe your reaction that our knowledge is too fragmentary?" Hardy answered,

> This is my view. In going through considerable toxicity literature as I did, for the December meeting, it is striking that nothing but harm has been reported as a lead effect either from industrial exposure or in experimental study. In contrast to many metals found in nature which are beneficial at certain levels, harmful at others, no useful effect has ever been ascribed to

lead. However, to make what is now known useful there is a world of work ahead. I don't think it is too difficult. It can be encompassed.[47]

In 1972 the known adverse metabolic and functional effects of increasing concentrations of lead on heme synthesis, the kidney, the nervous system, and other organ systems were reviewed by a special committee of the National Academy of Sciences. It is clear that lead interferes with the biosynthesis of heme at several enzymatic steps, with the utilization of iron, and in erythrocytes with globin synthesis. Other enzyme inhibitions may occur. It was suggested from in vitro studies that impaired aminolevulinic acid dehydrogenase (ALAD) activity in red blood cells is the earliest evidence of an adverse metabolic effect of environmental exposure to lead. Inhibition of ALAD activity in blood and increasing excretion of delta-aminolevulinic acid (ALA) in humans is associated with increasing concentrations of lead in blood. These effects may be the first indicators of subsequent clinical effects of lead poisoning and are referred to as *subclinical effects*.[48] A large body of experimental work is focused on the subclinical effects of lead. Discussion of this work is beyond the scope of this chapter.

OSHA LEAD STANDARD

In 1978 the Occupational Safety and Health Administration issued a permanent health standard for lead in the workplace. The standard sets limits on ambient concentrations of lead in workplace air and requires engineering controls, work practices, monitoring, and surveillance. It requires that workers be informed about the health effects of lead and methods of protection. Another feature of this standard, Medical Removal Protection, requires employers to measure workers' blood levels regularly; if the concentration of lead in the blood exceeds certain limits, the workers must be removed from exposure to lead until the level drops to an acceptable level. The workers' seniority status and wages must be maintained. OSHA allowed companies three to ten years to attain the limit of 50 micrograms per cubic meter through engineering controls.[49]

LEAD IN THE ENVIRONMENT

The environmental movement of the 1970s and mounting public concern made lead in the environment a public issue. As evidence of adverse health effects from low levels of lead exposure grew, a number of measures were taken in the United States to lower lead levels in the environment. Leaded gasoline, considered a major contributor to lead in the environment, has been largely phased out. The Environmental Protection Agency (EPA) established maximum contaminant levels for lead in drinking water.[50]

Recent environmental concerns center on lead in children, especially from exposure to lead paint. Lead poisoning from paint accounts for the majority of known cases of lead poisoning, especially in children in the United States and the United Kingdom.[51] In the United States a number of government agencies have acted in an effort to remove or control the hazard associated with lead paint. The Centers for Disease Control (CDC) recommends that screening for lead poisoning be included in health care programs for children and is currently reviewing its screening level for blood lead. The Consumer Product Safety Commission (CPSC) does not allow more than 0.06 percent lead in most paints.[52] The Department of Housing and Urban Development (HUD) has issued regulations requiring testing for and elimination of lead-based paint hazards in federally funded housing, housing rehabilitation programs, public housing, and Indian housing.[53]

SUMMARY

The significance of the history of plumbism up to this point is that an ancient disease once considered acceptable has been reevaluated and become a serious medical, human, and social problem. This change of focus is based on new scientific data that have changed our perceptions of the earlier manifestations of lead poisoning and created new attitudes toward the effects of lead poisoning on the health of industrial workers and the general population exposed to lead insult.

New uses of lead, new ability to make more accurate measurements, advances in biochemical science, and new attitudes toward the public health have all meshed to challenge the traditional concept of lead poisoning. In the dialogue initiated between those who see a threat and those who do not, the concept of health risk and even the concept of disease itself are undergoing redefinition.

References
1. Hunter, Donald. *The Diseases of Occupations*. London: The English Universities Press, 1957.
2. Lead Industries Association. *Facts about Lead in the Atmosphere*. New York: Lead Industrial Association, 1968.
3. Legge, Thomas M. and Kenneth W. Goadby. *Lead Poisoning and Absorption*. New York: Longmans, Green & Company, 1912.
4. Hunter, *Diseases of Occupations*, 235–242.
5. Agricola, Georgius. *De Re Metallica*. Trans. H. C. Hoover and L. H. Hoover. New York: Dover, 1950.

6. Taylor, Sherwood F. *A History of Industrial Chemistry*. New York: Abelard-Schuman, 1957.
7. Ibid., 83.
8. Hunter, *Diseases of Occupations*, 219.
9. Agricola, *De Re Metallica*, 4–5.
10. Rosen, George. *The History of Miner's Disease*. New York: Schumans, 1943. (quoting Paracelsus).
11. Ibid., p. 98 (quoting Martin Pansa).
12. Ibid., 99. (quoting Vernatti).
13. Ibid., 84.
14. Ramazzini, Bernadino. *Diseases of Workers,* New York: Hafner Publishing, 1964.
15. Ibid., 39.
16. McCord, Carey. "Lead and Lead Poisoning in Early America." *Industrial Hygiene and Surgery* 22 (September–December) 1953.
17. Ibid., 19.
18. Baker, George. *An Essay Concerning the Cause of Endemial Colic of Devonshire*. 1776; reprint, Delta Omega Society, 1958.
19. James, R. *A Medical Dictionary*. Vol. 3. London: T. Osborne, 1745.
20. Encyclopaedia Britannica. Edinburgh: A. Bell & Macfarouher, 1797.
21. Rees, Abraham. *The Encyclopedia or Universal Dictionary of Art, Sciences and Literature*. Vol. 20. London: Longman, Hurst, Rees, Orme & Brown, 1819.
22. Semmes, Thomas. *An Essay on the Effects of Lead*. Philadelphia: Carr & Smith, 1801.
23. Gray, J. B. "Colic Pictonum from the Medicinal Employment of Acetate of Lead." *Lancet* 1(1841–42): 123.
24. Kranzberg, Melvin and Carroll W. Purcell, eds. *Technology in Western Civilization*. Vol. 1. New York: Oxford University Press, 1967.
25. Kober, George M. and Emery Hayhurst. *Industrial Health*. Philadelphia: Blackistons Sons & Company, 1924.
26. Legge and Goadby, *Lead Poisoning*.
27. Hamilton, Alice, *Exploring the Dangerous Trades*. Boston: Little, Brown & Co., 1943.
28. Hatch, Theodore, *Industrial Hygiene Highlights*. Vol. 1. Pittsburgh: Industrial Hygiene Foundation, 1968.
29. Hearings Before a Subcommittee on Air and Water Pollution, Committee on Public Works, U.S. Senate, *Air Pollution*, 89th Congress, second session on S3112 and S3400, 1966,
30. Committee on Public Works, *Air Pollution*, 19.
31. Patty, Frank. *Industrial Hygiene and Toxicology*. New York: Interscience Publication, 1962.
31A. United States Public Health Service. *Survey of Lead in the Atmosphere of Three Urban Communities*. No. 999-AP12. Cincinnati: U.S. Public Health Service, 1965.

32. Chow, Tsaiwa and M. S. Johnstone. "Lead Isotopes in Gasoline and Aerosols of Los Angeles Basin." *Science* 147 (January 1965): 502.
33. Patterson, Claire. "Contaminated and Natural Lead Environments of Man." *Archives of Environmental Health* 2 (September 1965): 358.
34. Goldsmith, John R. and Alfred C. Hexter. "Respiratory Exposure to Lead: Epidemiological and Experimental Dose-Relationships." *Science* 158 (October 1967): 132–134.
35. Hardy, Harriet. "What Is the Status of Knowledge of the Toxic Effects of Lead on Identifiable Groups in the Population?" *Clinical Pharmacology and Therapeutics* 7 (November–December 1966): 713–722.
36. King, Barry G., A. F. Schaplowsky, and Edward B. McCabe. "Occupational Health and Child Lead Poisoning: Mutual Interests and Special Problems." *American Journal of Public Health* 62 (August 1972): 1056–1059.
37. *Pediatrics* 48 (September 1971): 464–468.
38. Oberle, Mark W. "Lead Poisoning: A Preventable Childhood Disease of the Slums." *Science* 165 (September 1969): 991–992.
39. Stopps, Gordon, Mary E. Maxfield, Martha McLaughlin, and Sidney Pell. *Lead Research: Current Medical Developments.* Pittsburgh: Industrial Hygiene Foundation, 1966.
40. Goldwater, J. Leonard and Walter A. Hoover. "An International Study of 'Normal' Levels of Lead in Blood and Urine." *Archives of Environmental Health* 15 (July 1967): 60–63.
41. United States Public Health Service. *Survey of Lead in the Atmosphere of Three Urban Communities.* No. 999-AP12. Cincinnati: U.S. Public Health Service, 1965.
42. Committee on Public Works, *Air Pollution,* 130.
43. U.S. Public Health Service, *Survey of Lead,* 19–20.
44. U.S. Public Health Service, *Symposium of Environmental Lead Contamination,* No. 1440, Washington: U.S. Government Printing Office, 1966.
45. Committee on Public Works, *Air Pollution,* 204.
46. Ibid., 325–326.
47. Ibid., 176–177.
48. National Academy of Sciences. *Lead, Airborne Lead in Perspective.* Washington: National Academy of Sciences, 1972.
49. Office of Technology Assessment. *Preventing Illness and Injury in the Workplace.* OTA-H-256 Washington: Congress of the United States, Office of Technology Assessment, 1985.
50. Agency for Toxic Substances and Disease Registry, *Toxicological Profile for Lead* Washington: U.S. Public Health Service, U.S. Environmental Protection Agency, 1990.
51. Lansdown, Richard and William Yule, eds. *Lead Toxicity, History and Environmental Impact.* Baltimore: The Johns Hopkins Press, 1986.
52. Agency for Toxic Substances and Disease Registry, *Toxicological Profile* 3–5.
53. Ibid., 151.

Chapter 5

Asbestos

Current public health consequences of poorly controlled utilization of asbestos in the past can be traced back, in part, to decisions made almost 50 years ago. Given the historical context of the 1940s and the desire to win World War II, the response to workplace health and safety hazards (in this case asbestos) during World War II is easily understood. However, a policy based on crisis, cursory risk assessment, and limited social commitment could not and did not last after the war. We now know that a lasting policy leading to worker health and safety was not the outcome of the hectic, ad hoc, wartime health and safety activities. If there is a lesson here, it is that occupational health policy based mainly on expediency is short-lived.

It is clear from this study that efforts to control disease associated with asbestos utilization during World War II depended on social and historical factors and that priorities reflected contemporary social values and necessities. The roots of the current asbestos disease problem that causes so much public concern can be traced, in part, to the shipyards of World War II, where extensive asbestos utilization, along with minimum controls and minor precautions, left a legacy of disease and death. Twenty years after

World War II, manifestations of asbestos diseases appeared with increasing frequency* and affected thousands of shipyard workers as well as other workers who had been exposed in the 1940s and during the postwar period, when injudicious applications multiplied the amount of asbestos utilized. Not until the 1960s did we begin to fully appreciate the legacy of disease and death that would result from the earlier high exposures to asbestos.[4, 5]

ASBESTOS: PROPERTIES, USES, AND HEALTH EFFECTS

Although asbestos had been used in small quantities for centuries, large-scale asbestos mining and commercial production started in the twentieth century and greatly accelerated during World War II. Figure 5-1 indicates asbestos consumption in the United States between the years 1912 and 1982. Asbestos consumption rose from less than 100,000 tons in 1912 to approximately 750,000 tons during World War II and 800,000 in the 1970s. The most rapid increase in consumption took place during World War II, when shipbuilding and ship repair used large amounts of asbestos. Since the beginning of the twentieth century, construction and manufacturing industries have used an estimated 30 million tons of asbestos, much of it introduced into the environment without significant precautions.

Industries that manufacture asbestos products or utilize them employ millions of people. In the mid-1970s, it was estimated that more than 37,000 persons were employed in manufacturing primary asbestos products, 300,000 worked in secondary asbestos industries, and millions worked in asbestos consumer industries, including 185,000 in shipyards and almost 2 million in automotive sales, service, and repair.[6]

At the present time it is well known and universally accepted that asbestos exposure can cause serious illness and death and that the major pathological effects of asbestos result from inhalation of fibers suspended in air.

---

* Three studies attempted to forecast future incidences of disease from past exposures of workers to asbestos in the United States. In the first study Peto and colleagues estimated that exposure prior to 1965 would cause approximately 37,500 mesothelioma deaths and 112,500 lung cancer deaths (at least three quarters have not yet occurred).[1] Nicholson and associates estimated that past exposures to asbestos would cause from 8,700 excess deaths of cancer per year in 1982 to almost 10,000 per year by 1990. After 1990, about 9,000 such deaths per year would occur until the twenty-first century, when the number would begin to decline.[2] Doll and Peto estimated that in 1978 perhaps 5 percent of the lung cancers in the United States could be attributed to occupational exposure to asbestos, an estimated 4,000 to 8,000 deaths per year.[3]

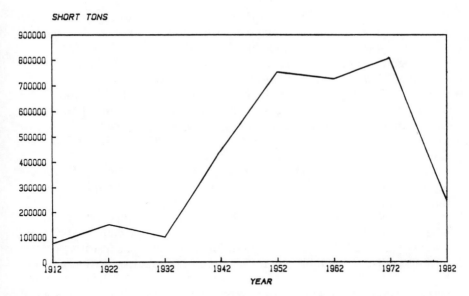

FIGURE 5-1.   Asbestos consumption in the United States between 1912 and 1982 (U.S. Department of the Interior, 1982).

RECOGNITION AND RESPONSE TO ASBESTOS-RELATED DISEASE BEFORE 1940

In order to understand events in the 1940s, when the transition from relatively small amounts of asbestos in industry to extremely large quantities took place, it is helpful to examine what was known about the health effects of asbestos to that time. Documentation of cases of asbestos-related disease began early in the twentieth century.[7, 8] At that time, lack of knowledge about risks associated with asbestos and the small amounts of the mineral used limited observations of and understanding about the relationship between asbestos and disease. The fibrotic lung disease did not receive a name until W. E. Cooke called it *asbestosis* in 1927.[9] In 1928 and 1929, the British government undertook an investigation of the condition of textile factory workers and reported to Parliament in 1930 that "inhalation of asbestos dust over a period of years results in the development of a serious type of fibrosis of the lungs." The commissioners recommended dust suppression. British Asbestos Industry Regulations followed in 1931.[10] Clinical reports in the United States also confirmed occurrence of asbestosis among asbestos workers. Reports of cases appeared in *The Journal of the American Medical Association*,[11] *American Journal of Public Health*,[12] and *Journal of Industrial Hygiene*.[13]

Recognition of adverse health effects as a result of exposure to asbestos continued. *Public Health Reports* published the results of a study undertaken by the Metropolitan Life Insurance Company.[14] The Dreessen study of asbestosis in the asbestos textile industry was published in 1938.[15] The first formal claims for compensation associated with asbestos disease in the United States were made in 1927, and in 1933 the Johns Manville Company settled all compensation claims out of court.[16]

A number of books on public health, medicine, and related subjects began to incorporate sections on industrial hazards and included asbestos dust among those hazards.[17–19]

By the 1940s, asbestos dust was identified as dangerous and unhealthy: inhalation of the dust over a period of years could cause asbestosis. Twenty years had elapsed between the initial reports of asbestos as a cause of fibrotic disease and the general acceptance of asbestos dust as a health hazard. Asbestosis is also generally associated with high levels of exposure, and in the early years of the twentieth century workers were exposed to heavy concentrations of asbestos dust. Years passed before a significant reduction of dust occurred. In the meantime, American and European industries used more and more asbestos and found new applications, seldom considering the health of workers.

The first reported case of association of asbestos dust and lung cancer appeared in 1935,[20] but the first rigorous epidemiological study appeared in 1955.[21] To complicate the matter, the latency period for cancer was often longer than that for fatal asbestosis, and lower levels of dust could cause cancer. Nevertheless, the foreshadowing of the tragedy to come had already begun in the 1930s, when British and American medical reports associated asbestos exposure with the development of lung cancer.

The Annual Report of the Chief Inspector of Factories for the Year 1947 reviewed all cases of workers in Great Britain known to have died with asbestosis since the first reported case in 1924. Thirteen percent had lung cancer; the expected incidence was 1 percent.[22] It took eight more years until Doll's epidemiological study, now a classic, documented the high risk of cancer in asbestos textile workers. Then, in 1964, the landmark Conference on Biological Effects of Asbestos, convened by the New York Academy of Sciences, resulted in a consensus among investigators of different nations that asbestos was a cause of lung cancer.[23]

The lengthy route from initial cognizance to confirmation to general acceptance of the association between asbestos and cancer once again, as in the case of asbestosis, took decades. Tragically, large numbers of workers would, in the meantime, become desperately ill, and many of them would die from a painful disease. Furthermore, in the 1960s it was becoming increasingly clear that the risk of disease was not confined solely to workers in mining and manufacturing but that it extended to those who used asbestos products as shipyard workers, insulation workers, and many others outside the primary or fixed-place industries.[24, 25]

Knowledge about health effects of asbestos was never secret. As early as the 1940s, a body of scientific and medical literature about fibrotic disease associated with the use of asbestos existed. It included, albeit rudimentary, measurement techniques and technology. Information could be found in textbooks, journals, and government publications. The literature was international.

Dust, a potential workplace hazard, led a number of experts to concentrate upon the problems of industrial dusts.[26] Apparently the academic, medical, industrial, and governmental communities, including the Public Health Service and the United States Navy, knew about the relationship between occupational disease and asbestos dust.

Although information appeared in professional journals, textbooks, and in the curriculum of a few universities, American society, as distinct from a relatively small core of public health and technical professionals, had not been widely sensitized to the idea that the work environment could contribute to future illness. Not until the 1970s and enactment of the watershed legislation that created the Occupational Safety and Health Administration did the general public become aware of and sustain concern about occupational safety and health issues.

WORLD WAR II: TRANSITION TO INCREASED UTILIZATION OF ASBESTOS

Thus, by the 1940s, while hazards associated with asbestos as a pneumoconiotic dust were generally recognized by a segment of the medical profession, and in some technical and scientific circles, public awareness of the risk associated with asbestos was limited. It took the advent of World War II and the concurrent industrial expansion that provided needed war materials to bring about a hitherto unprecedented increment in the amount of asbestos utilized and the number of people at risk of disease by virtue of such use.

Prior to World War II, the United States was ill equipped to fight, no less win, a war and urgently needed to stimulate war industries and production in order to build and equip its armed forces. In 1940, Congress voted large sums of money to build a two-ocean navy, as well as thousands of planes and equipment for an immense army. In 1941, Lend-Lease allowed the United States to make its shipyards available for repair and reconditioning of ships of any nation whose defense the president deemed vital to the interests of the United States. All this led to expansion of shipbuilding and repair, perhaps the most important item in a war economy that desperately needed to build and repair ships quickly.

In 1942 and 1943, German submarines took a heavy toll of Allied shipping. During a 15-month period, losses to submarines, air attacks, and marine casualties were close to 10 million tons. Ship production in the United States and Allied shipyards did not begin to replace these losses until the end of 1942.[27]

The United States Navy Department and the United States Maritime Commission administered shipbuilding activities. The Maritime Commission had been created in 1936 in an effort to modernize the Merchant Marine. The task facing the American shipbuilding industry—to build merchant ships faster than they were being sunk—continued throughout the war.

Maritime Commission programs were impressive: 5,601 vessels were delivered under Maritime Commission contracts. Additional merchant vessels, built in 1939 to 1945 for private companies and foreign governments, brought the total up to 5,777 vessels. Their construction consumed a total of about 25 million tons of carbon steel.[28] At the peak of employment in November 1943, it engaged the labor of 1,750,000 workers in shipyards: "No single item in the war economy was more important than shipbuilding, and nowhere was the need for speed more important."[29] Thus were priorities apportioned and the stage set.

When the United States Navy and the United States Maritime Commission became one of the largest industrial employers in America in the early 1940s, they had one major, overwhelming, and all-consuming priority: to build and repair enough ships to win the war. Along with the demand for increased production, unprecedented in the history of the United States, problems of shortages of materials and of qualified workers arose. Asbestos, a much needed insulating material used in shipbuilding, was one of the scarce substances.

Faced with shortages of labor, the Navy and Maritime Commission had to define and carry out a policy to keep workers safe, healthy, and productive. Along with the demand for increased production came the necessity to curtail accidents and health hazards that might slow down production. The known hazards to health included silica dust, welding fumes, solvents, lead, mercury, and asbestos. At the time, because the medical and industrial health community did not consider asbestos the most dangerous of the many hazardous materials, risks associated with lead, silica dust, and welding received far more attention. They were better understood and were believed to be more dangerous and widespread.

ORGANIZING AND DEVELOPING A
POLICY FOR SAFETY AND HEALTH

The Navy Department knew about industrial hazards as early as the 1920s. For example, in 1922 "Occupational Hazards and Diagnostic Signs: A Guide to Impairments to Be Looked for in Hazardous Occupations" was published in *United States Naval Medical Bulletin*.[30] A section on inorganic dusts described symptoms and conditions to look for and occupations that offered such exposure. The authors discussed four methods they considered effective to control dust inhalation: water to dampen the dust, exhaust systems to remove the dust, engineering designed to isolate the dust-producing process, and respirators. One occupational category the authors listed was asbestos workers.

In 1940, Captain Ernest M. Brown, USN, presented a paper at the Air Hygiene Foundation entitled "Industrial Hygiene and the Navy in National Defense" that listed 13 occupational hazards in United States Navy yards.[31] Brown included asbestosis among makers of pipe-insulating covers as one of two dust diseases. Silicosis was the other.

The immense increase in the number of ships built and repaired and the concurrent expansion of the work force that started in 1939 served to augment old problems related to health and safety. In February 1941, both Secretary of the Navy Frank Knox and Assistant Secretary of the Navy Ralph Bard sent memoranda to commandants and commanding officers of shore establishments. The subject of the Bard memo was "Safety and Health of Civil Employees."[32] The Knox memo referred to "Expansion of Safety and Health Programs in Naval Industrial Shore Establishments." Knox's memo stated that the navy had for many years pursued a safety program and an industrial hygiene program, but that in order to continue the work of the past and to "cope with the new problems constantly arising incident to technological advances, and the increased hazards due to the increasing number of civil employees incident to the existing emergency, it has been decided to expand the present safety and health organizations ashore." The memo provided "as soon as personnel was available" for the establishment of an industrial health office under the commandant or commanding officer in each naval district. Each office would employ a medical officer for industrial hygiene and medicine and a technical medical officer. The new officers would be trained in special courses already available.[33] Knox did not clarify the duties of these new officers.

The Bard memo noted the increase in severity and frequency of accidents that accompanied the growing number of employees. He requested high officials and supervisors to give their attention to "increased safety in working conditions" and to "indoctrinate all hands accordingly." Bard requested action to detect and correct unsafe conditions and practices and, if possible, to enlarge the safety engineering organization. In his list of items of importance, he included health hazards from dust.

In a February 1942 memo from Secretary Knox, angry because he believed that some commandants, production managers, and senior medical officers regarded the program to expand the safety and health organization as unprofitable, he stated that the program to expand the safety and health organization "shall be made a workable program." The memo reiterated the functions and duties of the medical officer for industrial hygiene and the need to utilize the skills of these officers.[34] Apparently the Navy Department was well aware of safety and health issues. The problem was one of how to initiate a workable and realistic occupational safety and health program during a time of crisis and of what priority should be given to this program.

The Navy turned to the Division of Industrial Hygiene in the newly created National Institutes of Health (a part of the Public Health Service) and asked it to provide industrial hygiene services. Because they had neither the funds nor the staff to take on such a task, the Public Health Service could not reply favorably to the Navy's request.[35] In the meantime, the work force continued to grow, and accidents in shipyards attracted attention. (Accidents are more easily enumerated and understood than health hazards.) Pressed by the need to maintain full operating forces to achieve maximum production, the Navy, in cooperation with the Maritime Commission, sponsored a joint project to survey the shipbuilding industry from the standpoint of accident prevention and control of industrial diseases. Daniel Ring, director of the Division of Shipyard Labor Relations, United States Maritime Commission, headed the project. The plan was to survey shipyards and then to take steps toward any needed corrective action that the surveys indicated. The Maritime Commission hired two consultants: Philip Drinker, a professor of industrial hygiene at Harvard University, to make health surveys of the shipyards; and John Roche, an industrial safety engineer from the National Safety Council, to survey accidents. Naval officers would assist in making the surveys, and all health and safety surveys were to be made available to the Navy as well as to the Maritime Commission.

Drinker and Roche completed their respective reports and made a series of recommendations. Drinker's recommendations included physical examinations for workers, proper use of respiratory equipment, periodic inspection of naval yards by physicians (so physicians could better understand the need for industrial safety and hygiene), additional physicians to work in navy yards, a uniform policy to handle ventilation problems, systematic labeling of solvents, minimization of the use of galvanized iron (to avoid metal fume fever), precautions to prevent lead poisoning, an end to open air sandblasting (because it created a silicosis risk to others than

the sandblasters), and a permanent medical and industrial hygiene program. Drinker wanted the industrial hygiene office to include the following
personnel: a full-time medical director from the shipbuilding industry, two
assistant medical directors from the Navy Medical Corps, and at least six
engineers trained in industrial hygiene.[36] Roche also made recommendations, based upon his survey.[37] Both had found numerous safety and health
hazards and an appalling lack of knowledge in navy shipyards about how to
protect workers from accidents and illnesses associated with shipbuilding.
The surveys revealed that accident and health problems were worse than
originally thought.[38]

Drinker presented the results of his health survey to the United States
Maritime Commission on October 20, 1942. In the report, Drinker stated
that in a west coast navy yard men were exposed to dangerous amounts of
dust from asbestos used in pipe covering and breechings. Drinker believed
the men exposed to asbestos should have periodic physical exams, but
they were not allowed to do so because of a labor contract. He also spoke
of respirators and said that "most of the yards on all coasts are either
careless, or ignorant, or both, in the case of respiratory protective equipment. This situation should be corrected and corrected promptly."[39] It is
clear from the Drinker report that asbestos was an industrial hazard known
to public health professionals and that methods existed for dealing with it
in the 1940s. It is also clear that the methods were not utilized throughout
the shipbuilding industry.

In November, Drinker and Roche met with a small group concerned
about what action to take. They recommended the preparation of minimum standards for safety and health in the shipyards. Rear Admiral Fisher
and Director Ring sent a memo to that effect to Assistant Secretary of the
Navy Bard and to the Commissioner of the Maritime Commission, Admiral Vickery.[40]

Drinker and Roche had already started writing the proposed minimum
standards (later called *minimum requirements*) they wished to see observed with respect to industrial health, hygiene, and safety measures in
the yards of all shipbuilding contractors for the Navy and the Maritime
Commission. The Navy and Maritime Commission began planning a
method for endorsement of the standards. Thus they proposed that the
following steps be taken in order to obtain cooperation: A two-day conference would be convened in December, at the joint invitation of Secretary
Knox and Admiral Land. Those invited would include safety engineers
from each plant, the head of the medical department from each plant, and
the labor representatives and one management representative from each
plant's labor-management committee. The purpose of holding the conference was to examine and discuss the proposed standards and then to adopt
them. After the conference did this, the standards would be promulgated

by the navy and the Maritime Commission, and a joint organization would be set up for advice, assistance, and "insuring compliance" with the standards, which would operate under the administrative direction of the Maritime Commission. The Maritime Commission planned to set up an organization for health with Philip Drinker in charge and available for consultation. Two medical officers detailed from the navy would serve under Drinker, and six officers from the Bureau of Medicine and Surgery would be integrated into the field operating forces of the Navy and Maritime Commission. Safety would be headed by Roche, and a safety expert would be assigned to each of the four regions.[41]

The scheduled conference became the vehicle to establish the United States Navy and Maritime Commission's policy on occupational safety and health in shipyards. Buried in the requirements was the issue of the health effects of asbestos. The plan of action set in motion in late 1942 was based upon a clear set of priorities. The minutes of that December conference reveal that the number one priority was efficient building and repairing of ships. Control of health and safety problems was a means to increase production.

THE CHICAGO CONFERENCE

The conference authorized by the United States Navy and the Maritime Commission met in Chicago on December 7–8, 1942 with 122 representatives attending. Participants included safety directors, safety engineers, physicians, members of labor-management committees, labor representatives, management representatives, representatives of the Navy and the Maritime Commission, representatives from insurance companies that covered compensation risks in contract shipyards, and representatives from a number of shipbuilding companies. The following discussion is based on stenographer's minutes of the 1942 Conference.[42]

United States Maritime Commission Director of Shipyard Labor Relations Daniel Ring presided over the first session on Monday morning, December 7. He said that the purpose of the meeting was to promote "standardized programs to develop at least minimum requirements for the two, safety and health, for the million people who will be in the shipyards after the first of the year." Ring stated the priorities of the Navy and the Maritime Commission: "We are building the ships to take the ammunition and supplies to the manpower of the nation, and it has been restricted and reduced in such a way that we must make sure that the farthest possible use must be made of everyone of us who are able to work on the home front." He said the conference was dedicated to "promoting the greatest individual efficiency in shipyards throughout the country. Our objective is based upon production because everything is subsidiary to that."

On the morning of the first day, Drinker and Roche presented the minimum standards for health and safety. On the second day, all participants met again to seek "a final agreement on the consensus of this conference with respect to those standards." The Navy Department and the Maritime Commission intended to promulgate the agreed-upon requirements immediately after the meeting. It was not clear what mechanisms would be utilized to enforce the requirements or if enforcement was ever intended. Indeed, before, during, and after the conference, the two terms *standards* and *requirements* were used as synonyms. Standards imply enforcement; requirements do not.

When Drinker summarized the requirements for industrial health for the meeting, he listed seven diseases likely to occur in shipyards. Asbestosis was one of the seven. Drinker said, "Asbestosis, that is not unlike silicosis in its effects, and we rather expect it to occur in shipyards because we have seen asbestos being handled in installation work with little or no precautions." He also said, "Asbestosis, silicosis, poisoning from solvent vapors and lead cannot be controlled satisfactorily unless examinations of men are enforced." During the afternoon session, Drinker again referred to asbestosis: ". . . danger is present in some yards and should be recognized."

However, asbestosis was only one of a number of occupational diseases that Drinker pointed out. Flash burns, foreign bodies in the eye, lead poisoning, solvent vapors, zinc fume fever from welding galvanized metals, fibrous glass, and silicosis were some other problems discussed. Record keeping, preplacement examinations, periodic physical examinations for workers who worked with substances hazardous to themselves as well as to others, and the responsibilities of the medical department were also discussed. Other agenda items included employment of women, disposal of waste, cafeterias, and respiratory protection. In the hundreds of pages of transcripts of the meeting, there are only a few references to asbestos. Both from the transcripts of the meeting and from the document that followed the meeting, it is clear that Drinker tried to apply, as best he could, the concepts of industrial hygiene of the period.

At the conclusion of the conference, the delegates recommended adoption of a set of "minimum" requirements for safety and industrial health in contract shipyards.

MINIMUM REQUIREMENTS

On January 20, 1943, the United States Navy approved the minimum requirements, and on February 9, 1943, the United States Maritime Commission did the same. Both then circulated the requirements entitled "Minimum Requirements for Safety and Industrial Health in Contract

Shipyards'' to all contractors constructing ships for the United States Navy and the United States Maritime Commission, as well as to the chiefs of all bureaus, commandants and commanding officers, naval shore establishments, and naval supervisors of shipbuilding.[43] The document, signed by Secretary of the Navy Frank Knox and Chairman of the United States Maritime Commission E. S. Land, stated policy and priorities in the following manner.

> To All Contractors Constructing Ships for The United States Navy and Maritime Commission:
>
> As a result of the national conference on safety and health in shipyards holding contracts with The United States Maritime Commission, conducted under the auspices of these agencies in Chicago December 7 and 8, 1942, a unanimous agreement was reached upon the minimum standards which have been approved by the Navy Department and United States Maritime contracts with the two agencies. These standards represent a specialized study based upon a fact finding survey on all coasts by experts in that field. They have received the unanimous concurrence of the representatives of the medical and safety departments and of labor-management committees from shipyards on all coasts.
>
> The necessity for conserving manpower and promoting physical welfare, health and safety of what shortly will amount to one million workers in shipyards requires that careful observance of standards for the prevention of accidents and protection of health be accorded. Aside from the weight which must be given humanitarian considerations, it is simply good common sense that as much care and attention be given to protecting the human factors in the war production program as is given machines.
>
> Under the administrative direction of the Maritime Commission, safety and industrial health consultants will be made available in all regions wherein shipyards holding contracts with the Navy and the Commission are located.
>
> Each contractor is hereby given notice that the Navy Department and the Maritime Commission will expect full and complete compliance with the minimum Standards which bear the approval of The Navy Department and The Maritime Commission, and each is requested to give full cooperation to the consultants on health and safety who will be charged with the coordination and supervision of the safety and health programs of the two agencies.
>
> The cumulative restriction of manpower makes speedy attention and comprehensive action in respect to the subject matter hereof of vital importance.[44]

The circulated document contained 12 pages of minimum requirements for health. Section 13.7, less than half a page long, covered the asbestos hazard:

13.7 Asbestosis

a. Sources: In general any job in which asbestos dust is breathed. For example:

Job:	When material is:
Handling	Asbestos
Sawing	Asbestos mixtures
Cutting	
Molding	
Welding	
and rod	
salvage	

b. Job can be done safely with:
  1. Segregation of dusty work and,
  2. (a) Special ventilation: Hoods enclosing the working process and having linear air velocities at all openings of 100 feet per minute, or
     (b) Wearing special respirators
  3. Periodic medical examinations.[45]

In March Secretary of the Navy Knox sent a memorandum to all chiefs of bureaus, commandants, and commanding officers, Naval Shore Establishment, and to naval supervisors of shipbuilding on the subject of minimum requirements for safety and industrial health. He included copies of *Minimum Requirements for Safety and Health in Contract Shipyards* and pointed out that attention should be paid to the cooperation and compliance expected and that the minimum requirements applied to all private shipyards having either Navy or Maritime Commission contracts. But the second paragraph of the memo clearly stated that the "Requirements" were supplied to the Naval Shore Establishment as reference material only and should not take precedence over conflicting navy instructions. Knox then instructed supervisors of shipbuilding to cooperate with representatives of the Maritime Commission assigned to the project, that is, Drinker and Roche.[46] Drinker and Roche were consultants; they did not have power to enforce the minimum requirements. Both men reported to Daniel Ring, director of shipyard labor relations of the Maritime Commission.

Offices were established in regional stations at Philadelphia, New Orleans, Oakland, and Chicago. The Navy assigned trained officers, physicians, and industrial hygienists to Drinker's staff and civilian engineers to Roche's staff. Drinker, Roche, and their assistants were responsible for assisting the Navy and the Maritime Commission to put the "Minimum Requirements" into effect in the shipyards. Together with their assistants, Drinker and Roche could merely make recommendations. They acted only in an advisory capacity and lacked power to compel navy and Maritime Commission contractors to comply with the provisions of the minimum requirements. According to Drinker, the document served as a guide for health and safety control in contract yards.[47] Reporting and inspections, which began in early 1943 and continued through V-J Day, were gradually cut down and ended entirely on 1 October 1945.[48]

It is difficult to know the extent of the impact, if any, of these unenforceable, short-lived requirements on health and safety between 1943 and 1945. Future cases of asbestosis, lung cancer, and mesothelioma among shipyard workers would clarify one part of that question. Apparently the short-lived requirements did little to control asbestos use or to make it safe.

CONCLUSIONS

In retrospect, the asbestos experience during World War II conforms to the shortcomings of earlier societal perceptions about occupational disease. We now know that the limited control of occupational exposures to asbestos during the war did not prevent future asbestos disease. We also know that asbestos, once related only to the workplace, would later be recognized as a hazard to the general public.

By the late 1960s, the social consequences of indiscriminate utilization of asbestos, coupled with tardy recognition and dilatory response to workplace environmental factors as determinants of health, were apparent. While the general public slowly reacted to the unfolding information about health risks associated with asbestos, the "magic mineral" had been taking its toll in seriously debilitating disease and painful death.

Early efforts to control asbestos depended upon social and historical factors, as well as upon science and technology, much as they do today. Scrutiny of past actions offers insights into priority setting and policy determinations regarding the utilization of valuable but toxic materials. Three issues with lessons from past experiences and implications for future policy emerge from this historical perspective: (1) priority setting and allocation of limited resources, (2) utilization of potentially toxic materials with inherent hazards that are suspected but not clearly understood or that are completely unknown, and (3) assumption of risk by an employee or the community.

The first issue, that of priorities and allocation of resources, had tremendous impact on shipyard health and safety decisions during World War II. A profusion of problems had developed from increased production and from the growing size of the labor force, which was composed of large numbers of inexperienced workers who labored in an extremely hazardous industry. The rationale of the Navy and the Maritime Commission was that, in order to maintain a productive work force, workers must be kept as safe and healthy as possible. The navy and the Maritime Commission developed and implemented a health and safety policy based on the need to quickly and efficiently build and repair as many ships as possible. Priorities were set in response to a crisis produced by the war and to the pressure of ship production despite shortages of skilled labor and materials.

If one uses the vocabulary of the 1980s, the Navy and the Maritime Commission based wartime health and safety measures on a cursory preliminary risk assessment, Roche and Drinker's reports of shipyard conditions. "Minimum Requirements for Safety and Industrial Health" reflected contemporary control methods, medical information, and industrial health concepts. The role of the consultants and the lack of any enforcement provision reflected contemporary societal values. Policy was not determined by humanitarian goals; but by the overwhelming, all-consuming priority, to build and repair ships in order to win the war.

The policy, based on response to crisis, neither lasted after the war nor had a positive impact upon future safe use of asbestos or other toxic materials. In fact, the war stimulated increased utilization of asbestos. If there is a lesson here about setting priorities, perhaps it is that it is difficult to set appropriate, enduring policies based on response to crisis.

The second issue, that of utilization of potentially toxic materials with inherent hazards (which may be demonstrable, not clearly understood, or even unknown), is very much a part of the asbestos legacy. In 1940, asbestos dust was a known industrial hazard. Methods existed to control asbestos dust, but the extent of the full range of known and potential dangers remained unclear, and the implications of unrestricted use were unexplored. Misunderstanding, the gap between identification and acceptance, and, in some cases, disregard for what was already known about hazards associated with asbestos, and wartime needs all allowed for unrestricted, minimally controlled asbestos utilization during World War II.

This situation antedates and frames but does not answer another major societal question: Why did it take so long *after* the war to control asbestos and other dangerous materials? It was not until the 1970s and passage of the Toxic Substances Control Act (TSCA) that a general United States policy for utilization of potentially toxic materials was articulated. TSCA called for premanufacture notice by manufacturers of the potentially harmful chemicals brought into the market and placed the burden of responsibility on those who would profit from the sale of a toxic material. In the 1940s toxic materials could be marketed without description of their hazardous properties. Knowledge of individual chemical toxicity was proprietary and only painstakingly gained by the user. Relatively few Material

Safety Data Sheets were available, and they were often incomplete, particularly with regard to toxicity data.

The Occupational Safety and Health Administration "Hazard Communication Rule" is one culmination of efforts to have chemical manufacturers document the potential hazard of chemicals. The rule covers only the manufacturing sector of the economy. Thirty states or localities have promulgated hazard communication or "right-to-know" rules for the workplace. These are, in general, broader in scope than the federal rule. In retrospect, asbestos is a unique material, and the constant reiteration of this fact could have diverted attention from the potential hazards of all the other toxic chemicals used in a technological society.

The third issue, assumption of risk by an employee or the community when neither is informed as to the risk, was not seriously questioned until the end of the 1960s. Public concern about health effects of asbestos heightened in the 1960s as evidence that asbestos could cause cancer became public knowledge. The issue of manufacturer divulgence of hazard to the user, mentioned previously, is separate from the employee right to know. Prior to the 1970s, the employer could have been informed of hazards on the job and in many cases was, but did not share this knowledge with employees. In World War II shipyards, the United States Navy and the Maritime Commission knew about asbestos toxicity. Enlisted men and employees did not share this knowledge. Information about asbestos toxicity had not yet become public knowledge, nor was the general public fully cognizant of the fact that a relationship existed between conditions in the workplace and the safety and health of the worker.

A word about historical perspective. A sense of tragedy pervades the study of asbestos utilization during World War II, first, because we now know the outcome of the intense utilization of asbestos in the 1940s. And, second, given the historical context of the need to win the war at any cost, asbestos might have been heavily utilized whether or not its toxic properties were known or future incidence of disease had been predictable, although more precautions would likely have been taken. It is clear that the state of scientific and technological knowledge is not the only determinant of occupational safety and health policy. Historical context and social factors also affect those decisions.

Although the material presented here provides insight into the policy affecting asbestos utilization during World War II, it does not address a major unanswered question: Why was asbestos utilization allowed to expand in a relatively uncontrolled manner after the war?

References
1. Peto, J., B. E. Henderson, and M. C. Pike. "Trends in Mesothelioma Incidence in the United States and the Forecast Epidemic due to Asbestos Expo-

sure during World War II." In R. Peto and M. Schneiderman, eds., *Banbury Report 9: Quantification of Occupational Cancer.* Cold Spring Harbor, NY: Cold Spring Harbor Laboratory, 1981.

2. Nicholson, W. J., G. Perkel, I. J. Selikoff, and H. Seidman. "Cancer from Occupational Asbestos Exposure: Projections 1980–2000." In R. Peto, and M. Schneiderman, eds., *Banbury Report 9: Quantification of Occupational Cancer.* Cold Spring Harbor, NY: Cold Spring Laboratory, 1981.

3. Doll, R. and R. Peto, "The Causes of Cancer: Quantitative Estimates of Avoidable Risks of Cancer in the United States Today." *Journal of the National Cancer Institute* 66(1981): 1191–1308.

4. Selikoff, I. J., J. Churg, and E. C. Hammond, "Asbestos Exposure and Neoplasia," *Journal of the American Medical Association* 188(1964): 22–26.

5. Selikoff, I. J., and J. Churg. "Biological Effects of Asbestos." *Annals of the New York Academy of Sciences* 132 (1965).

6. National Cancer Institute. *Asbestos: An Information Resource.* DHEW Publication 79-1681 Washington: Department of Health, Education and Welfare, 1978.

7. Auribault, M. "Note sur l'hygiene et la securite des ouvriers dans les filatures et tissages d'amiante." *Bull de l'Inspection due Travail* 14(1906): 126.

8. Murray, H. M. Statement before the committee in the minutes of evidence. *Report of the Departmental Committee on Compensation for Industrial Disease.* London: His Majesty's Stationery Office, 1907.

9. Cooke, W. E. "Pulmonary asbestosis." *British Medical Journal* 2(1927): 1024–1025.

10. Legge, T. *Industrial Maladies.* London: Oxford University Press. 1934.

11. Lynch, K. M. and W. A. Smith, "Pulmonary Asbestosis. III. Carcinoma of Lung in Asbestos-Silicosis." *American Journal of Cancer* 14(1935): 56–64.

12. Donnelley, J. "Pulmonary Asbestosis." *American Journal of Public Health* 23(1933): 1275–1281.

13. Ellman, P. "Pulmonary Asbestosis: Its Clinical, Radiological and Pathological Features and Associated Risk of Tuberculosis Infection." *Journal of Industrial Hygiene* 15(1933): 165–183.

14. Lanza A. J., W. J. McConnell, and J. W. Fehnel, "Effects of the Inhalation of Asbestos Dust on the Lungs of Asbestos Workers." *Public Health Report* 50(1935): 1–12.

15. Dreessen, C. W. *A Study of Asbestos in the Asbestos Textile Industry,* U.S. Public Health Bulletin 241 Washington: Government Printing Office, 1938.

16. *Report of the Royal Commission on Matters of Health and Safety Arising from the Use of Asbestos in Ontario.* Vol. 1. Toronto: Ontario Ministry of the Attorney General, 1984.

17. Lanza, A. J. ed. *Silicosis and Asbestosis.* New York: Oxford University Press, 1938.

18. Clark, W. I. and P. Drinker, *Industrial Medicine.* New York: National Medical Book Co., 1935.

19. Rosenau, M. J. *Preventive Medicine and Hygiene.* New York: Appleton-Century-Crofts, 1935.

20. Lynch, K. M. and W. A. Smith, "Asbestos Bodies in Sputum and Lung." *Journal of the American Medical Association* 95(1930): 659–661.
21. Doll, R. "Mortality from Lung Cancer in Asbestos Workers." *British Journal of Industrial Medicine* 12(1955): 81–86.
22. Merewether, E.R.A. *Annual Report of the Chief Inspector of Factories for the Year 1947.* London: His Majesty's Stationery Office, 1947.
23. Selikoff, I. J., J. Churg, and E. C. Hammond. "The Occurrence of Asbestosis among Insulation Workers in the United States." *Annals of the New York Academy of Sciences* 132(1965): 139–155.
24. Selikoff, Churg, and Hammond, "Asbestos Exposure."
25. Harries, P. G. "Asbestos Hazards in Naval Dockyards." *Annals of Occupational Hygiene* 2(1968): 135–145.
26. Drinker, P. and T. Hatch, *Industrial Dust.* New York: McGraw-Hill Book Co., 1936.
27. Morrison, S. E. and H. S. Commager, *The Growth of the American Republic,* Vol. 2. New York: Oxford University Press, 1956.
28. Lane, F. *Ships for Victory.* Baltimore: The Johns Hopkins Press, 1951.
29. Morrison and Commager, *Growth of the American Republic.*
30. Dublin, L. and P. Leiboff, "Occupational Hazards and Diagnostic Signs: A Guide to Impairments to Be Looked for in Hazardous Occupations." *United States Naval Medical Bulletin* 17(1922): 893–914.
31. Brown, Ernest M., "Industrial Hygiene and the Navy in National Defense." *War Medicine* 1(1941): 3–14.
32. Memorandum from Ralph A. Bard, Assistant Secretary of the Navy (21 February 1941): "To Commandants and Commanding Officers, Shore Establishments. Subject: Safety and Health of Civil Employees." Hygiene, Commandant Correspondence. Record Group 181, National Archives, Boston.
33. Memorandum from Frank Knox, Secretary of the Navy (24 February 1941): "To Chiefs of All Bureaus, Commandants and Commanding Officers, Naval Shore Establishment: Naval Supervisors of Shipbuilding: Subject: Expansion of Safety and Industrial Health Programs in Naval Industrial Shore Establishments." Shore Establishments Division, Shipyard Labor Relations. Record Group 80, National Archives, Washington, D.C.
34. Memorandum from Frank Knox, Secretary of the Navy (23 February 1942): "To Chiefs of All Bureaus, Commandants and Commanding Officers, Naval Shore Establishment: Naval Supervisors of Shipbuilding: Subject: Industrial Health Program, Duties of Medical Officers for Industrial Hygiene." Shore Establishments Division, Shipyard Labor Relations. Record Group 80, National Archives, Washington, D.C..
35. Doyle, H. N. "The Federal Industrial Hygiene Agency: A History of the Division of Occupational Health, United States Public Health Service." Paper delivered at the American Conferences of Governmental Industrial Hygienists, Cincinnati, n.d.
36. Drinker, Philip. "Report to the Maritime Commission." Health and Safety, Shipyard Labor Relations. Record Group 178. 1942. National Archives, Washington, D.C.

37. Roche, John M. "Report to the Maritime Commission." Health and Safety, Shipyard Labor Relations. Record Group 178. 1942. National Archives, Washington, D.C.
38. Drinker, Philip. "The Health and Safety Program of the United States Navy." *Journal of the American Medical Association* 121(1943): 822–823.
39. Drinker, Philip. Talk before the Maritime Commission. Health and Safety, Shipyard Labor Relations. Record Group 178. 1942. National Archives, Washington, D.C.
40. Memorandum from Rear Admiral Fisher and Daniel S. Ring (5 November 1942): "To Ralph Bard, Assistant Secretary of the Navy and Admiral H. L. Vickery, Commissioner, U.S. Maritime Commission: Subject: Plans for Safety and Health Programs." Shore Establishments Division. Shipyard Labor Relations. Record Group 80, National Archives, Washington, D.C.
41. Memorandum from Fisher and Ring.
42. Stenographer's minutes before the United States Maritime Commission meeting in regard to minimum requirements for industrial health and safety. U.S. Department of Transportation, Maritime Administration: Chicago, 1942.
43. Memorandum from Frank Knox, Secretary of the Navy (4 March 1943): "To Chiefs of All Bureaus, Commandants and Commanding Officers, Naval Shore Establishment and Naval Supervisors of Shipbuilding. Subject: Minimum Requirements for Industrial Health and Safety." Shore Establishments Division, Shipyard Labor Relations. Record Group 178, National Archives, Washington D.C.
44. *Minimum Requirements for Safety and Health in Contract Shipyards*. The National Archives, Washington: U.S. Government Printing Office, 1943.
45. *Minimum Requirements*.
46. Memorandum from Frank Knox, Secretary of the Navy (5 May 1943): "To Chiefs of All Bureaus, Commandants and Commanding Officers, Naval Shore Establishment and Naval Supervisors of Shipbuilding. Subject: Joint Navy–Maritime Commission Project with Respect to Safety and Industrial Health in Private Shipyards Having Navy Contracts." Shore Establishments Division, Shipyard Labor Relations. Record Group 178, National Archives, Washington, D.C.
47. Drinker, "Health and Safety Program."
48. Drinker, Philip. "Health and Safety in Contract Shipyards during the War." *Occupational Medicine* 3(1947): 335–343.
49. United States Department of the Interior. *Bureau of Mines Mineral Yearbook: Asbestos*. Washington: U.S. Department of the Interior, 1982.

Chapter 6

Free Silica (SiO$_2$)

Silicosis, one of civilization's oldest known occupational diseases, has at various times been named *dust consumption, ganister disease, grinders' asthma, grinders' consumption, grinders' rot, grit consumption, masons' disease, miners' asthma, miners' phthisis, potters' rot, sewer disease, stone masons' disease, chalicosis,* and *schistosis.* Knowledge of the association between inhalation of dust and its ill effect on an individual's health has a long history, which started in ancient times with the recognition that a connection existed between inhalation of mine dust and the diseases that affected the lungs of miners.

Silicosis, although "ancient," remains with us and still represents an important occupational health problem. In the 1970s the National Institute for Occupational Safety and Health (NIOSH) estimated the number of workers employed in industries with potential exposure to silica dust[1] (Table 6-1). The NIOSH survey provided, for the first time in the United States, a database that presented industry-specific estimates of the number of workers exposed to a workplace hazard.

In 1980 the United States Department of Labor submitted an interim report to Congress on occupational disease that estimated the number of workers exposed to silica dust and the number of silicosis cases.[2] OSHA estimated 1 million workers currently exposed to free silica in metal mining, rock quarrying, glass, ceramic, cement, brick production, and abrasive blasting. The estimates did not include 700,000 heavy construction workers, 2.5 million agricultural workers, 600,000 chemical and allied products workers, and others with varying levels of exposure to free silica. Although the exact percentage of workers exposed to silica and the amount of exposure were not known (because of inadequate medical and

TABLE 6-1.   Employment in U.S. Industries Having Potential Exposure to Free Silica

Industry	Exposed Employees (1970)
Metal mining	76,000
Coal mining	125,000
Nonmetallic minerals (except fuels)	95,000
Stone, clay, and glass products	507,000
Iron and steel foundries	188,000
Nonferrous foundries	69,000
Total	1,060,000

This table omits large numbers of workers employed in the production of chemicals and allied industries, workers employed in heavy construction, and many others who may be exposed to free silica.

SOURCE: Reference 1.

environmental monitoring and unclear job classifications), it was estimated that 59,000 exposed persons would ultimately suffer from all stages of silicosis. Table 6-2 indicates estimates of the number of workers exposed to silica dust in certain industries and the number of silicosis cases.

Although NIOSH recommended criteria for a permanent standard to protect workers from exposure to the hazard of silica dust to OSHA in the 1970s, a permanent standard for silica dust has not yet been promulgated. Currently, permissible limits for airborne silica in the workplace are contained in OSHA and MSHA (Mine Safety and Health Administration) regulations. These regulations do not contain provisions for physical examinations, engineering controls, record keeping, or protective equipment.

World Health Organization statistics in the 1970s indicated that "in many industrialized countries silicosis is still a serious occupational disease, in spite of the fact that a reduction in incidence has been observed in mining in some countries, thanks to technological success with dust control measures."[3] Table 6-3 displays statistics for the number of cases of silicosis reported in both industrialized and developing nations in the 1970s. Silicosis is still present in industrialized nations; in developing countries the problem is often more serious. Diagnosed cases reflect earlier conditions of exposure, but insufficient data are available to establish

TABLE 6-2.   Workers Exposed to Silica Dust and Silicosis Cases

Industry	Current Number of Exposed Workers*	Estimated Number of Silicosis Cases
Mining		
Metal	24,000	816
Coal	126,000	630
Nonmetal	7,000	56
Quarry		
Granite	4,000	400
Sand and gravel	40,000	4,000
Stone, clay, and glass products	511,000	31,500
Foundries		
Iron and steel	192,000	16,700
Nonferrous	75,000	
Abrasive Blasting		5,000
Total	1,057,000	59,102

* Source: Bureau of Labor Statistics *Employment and Earnings*, February 1978; Mine Safety & Health Administration.

trends. Therefore, only future statistics will reflect current environmental conditions in these nations.

This chapter traces the extension of knowledge about silicosis in order to provide historical perspective to help understand the disease and measures taken to prevent or control the spread of silicosis. In antiquity it affected only miners, but silicosis became widespread and concomitant with the growth of technology and industrial society. Until the twentieth

TABLE 6-3.   Reported Cases of Silicosis

	Number of Exposed Workers in Mining and Quarrying*	Reported Cases	Year
New South Wales (Australia)	65,000	2,448	1974
Spain	4,506,000	1,656	1973
German Democratic Republic (coal mines only)	191,000	581	1973
Poland	457,000	1,008	1973
Bolivia (tin mines only)	95,000†	3,498	1971
Columbia	108,500	280‡	1972

* Figures from *Yearbook of Industrial Statistics*, Vol. 1. (New York: Department of Economic and Social Affairs, Statistical Office of the United Nations, 1978).
† Based on 5 percent of 1.9×10^6 work force in mining and quarrying (*Statistical Abstracts of Latin America*, 1977).
‡ Of 1,991 workers exposed to dust.
Source: Reference 3.

century, silicosis was widely misunderstood and inadequately defined. Then a significant increase in workers at risk generated new interest, new information, and new concepts of control and prevention. The problem of controlling or eradicating this severe, irreversible, disabling, but also preventable disease now, as in the past, demands technical, social, economic, and political solutions.

NATURE OF THE DISEASE

Silicosis, "a condition of massive fibrosis of the lungs marked by shortness of breath and resulting from prolonged inhalation of silica dust,"[4] is one of the pneumoconioses, a group of lung diseases that result from inhalation of excessive amounts of dust in certain occupations. Silicosis describes only the condition caused by inhaling uncombined or free silica (SiO$_2$); only dust containing free silica can cause silicosis. The disease can develop rapidly, but in general it is a chronic or slowly developing disease.

Silicosis is usually divided into three stages: slight, moderate, and severe. The onset of symptoms is marked by dyspnea or shortness of breath on exertion, at first slight and later increasing in severity. Shortness of breath remains an important symptom throughout the illness. A slight and unproductive cough may also be present. In this first or slight stage, the patient's general condition is unimpaired, physical signs in the chest are slight, and diminished expansion is scarcely present. Dullness can rarely be demonstrated, and there is no alteration of breath sounds or added sounds. Impairment of working capacity may be slight or absent. In the second or moderate stage, dyspnea and cough become established, and other physical signs begin to appear, such as diminished expansion of chest and patchy dullness, sometimes with bronchial breath sounds. There is impairment of working capacity. In the third stage, dyspnea leads to incapacity. Right heart hypertrophy and then failure may supervene. In the first or slight stage, the radiograph shows the presence of discrete modular shadows. In the second or moderate stage, both lung fields are wholly occupied by modular shadows, and there is some coalescence to form more or less dense opacities. In the third or severe stage, radiographs indicate massive consolidation.[5] Pulmonary tuberculosis frequently accompanies silicosis; it may be present at any stage of the disease.

The following description of silicosis, extracted from a primer prepared for workers, graphically illustrates the disease's progress.

> The main symptom is shortness of breath, at first occurring only during
> physical activity, but soon appearing after less and less exertion, until
> eventually the victim is short of breath, even at rest. This is caused by many

small round lung scars that develop from irritation by silica dust. These hard inelastic scars—just like those on the skin that result from an operation—make the lungs stiff, so that it takes more work to inflate them with air. The scars also thicken the walls of the air sacs, blocking the transfer of oxygen into the blood; low blood oxygen is a characteristic finding in silicosis. The area surrounding each scar becomes stretched and distorted, breaking down the normally tiny, delicate air sacs so that they form larger, thicker-walled sacs, a form of localized emphysema. Further reaction to the silica may cause scars to join into larger scars; some may occupy the entire lung. This process, progressive massive fibrosis, is frequently accompanied by increasing susceptibility to tuberculosis and other infections. Finally, the heart, which must pump blood through these stiff inelastic lungs, becomes weakened and enlarged and fails to pump effectively.[6]

OCCURRENCE OF FREE SILICA

Silica, the cause of silicosis, is the most abundant constituent of the minerals and rocks that make up the earth's crust. It occurs in two forms: free and combined. The most common form of free silica is quartz, a hard mineral and an abundant constituent of granite, schist, and other rocks and the chief component of sandstone and quartzite. Numerous ores are in veins consisting nearly wholly of quartz. Free silica exists in nature as opal, flint, tridymite, cristobalite, siliceous glass, or vitreous silica.

Because the earth's crust contains such large amounts of silica, mining and tunnel building are occupations closely associated with the silica hazard. The sandstone industry, quarrying, the granite industry, the pottery industry, slate quarrying and dressing, grinding of metals, iron and steel foundries, silica milling, flint crushing, the manufacture of abrasives, and the manufacture of abrasive soaps are also occupations that expose workers to silicosis. American industries consume millions of tons of quartz annually. Sand, an essential ingredient of concrete and mortar, is quartz, as are sandstone and quartzite, which are used as building stones. Crushed sandstone and quartzite are used for road and railway construction, roofing granules, and riprap erosion-control linings for river channels. Sand is used as a sandpaper abrasive and in sandblasting; for polishing and cutting glass, stone, and metal; and for providing traction on stairs, streets, and rails. Quartz is used in refractory products such as insulating and firebricks, foundry molds, and electrical insulators. Quartz is an ingredient in glass and porcelain manufacture. High-quality quartz is fused to make premium grades of chemical and optical glass. Fibers of vitreous silica are used in precision instruments such as balances, galvanometers, and gravimeters, and large tonnages of quartz of various qualities are used as raw materials in processes in which silica is not the final product: production of water glass or sodium silicate, various sols—very fine dispersions of solids in liquids—used in hydrophobic coatings, organic silicates and silicones, silicon carbide, silicon metal, and smelting, flux, and alloying in metallurgy.[7]

Today there is a suspected relation to cancer risk because of both epidemiologic and experimental data bearing on the relationships between silica exposure or silicosis and the development of lung cancer. The International Agency for Research on Cancer (IARC) in its 1987 monograph *Evaluation of the Carcinogenic Risk of Chemicals to Humans: Silica and Silicates* came to the following conclusions:

> There is *sufficient evidence* for carcinogenesis of crystalline silica to experimental animals. There is *inadequate evidence* for the carcinogenesis of amorphous silica to experimental animals. There is *limited evidence* for the carcinogenesis of crystalline silica to humans. There is *inadequate evidence* for carcinogenesis of amorphous silica to humans.[8]

Needless to say, the IARC classification of crystalline free silica as an animal carcinogen and a possible human carcinogen is a very controversial subject in the scientific community.

EARLY RECOGNITION OF THE DISEASE

Silicosis has always been associated with the history of mining, which began in Neolithic times. Miners were exposed to dusts that caused disease. It is possible that Neolithic miners who dug flint from the earth were the first people to suffer from silicosis. A few references to dust diseases exist in ancient medical literature, beginning with Hippocrates. He noted the connection between dust and disease in a metal miner who breathed with difficulty and had other symptoms similar to those found in silicosis. Pliny described devices used by miners to prevent inhalation of "fatal dust," and Celsus wrote, "By far the most terrible form of emaciation is that which the Greeks call phthisis. It spreads to the lung. On top of this, ulceration occurs and a slow fever which at times disappears and at other times reappears."[9] The association between mining, dust, and disease had been made. George Rosen noted that little medical interest in miners' diseases existed in the world of antiquity because of the miner's inferior status as a servile laborer. Rosen interpreted this as an illustration of the social indifference with which miners were viewed.[10] There are no contributions to the subject of miners or dust diseases to be found during the medieval ages, but in the sixteenth century Agricola and Paracelsus presented relevant information.

In 1556, Agricola published his classic encyclopedic treatise on the mining and metal industries. It included a description of the ills and accidents that afflicted miners.

> It remains for me to speak of the ailments and accidents of miners, and the methods by which they can guard against these. . . . On the other hand, some mines are so dry that they are entirely devoid of water, and this dryness causes the workmen even greater harm, for the dust which is stirred and beaten up by digging penetrates into the windpipe and lungs, and produces difficulty in breathing, and the disease which the Greeks call asthma. If the dust has corrosive qualities, it eats away the lungs and implants consumption in the body; hence, in the mines of the Carpathian Mountains women were found who have married seven husbands, all of whom this terrible consumption has carried off into premature death.[11]

Paracelsus described the bergsucht or miners' phthisis and traced it to inhalation of metal vapors and not to, as many believed, mountain spirits who punished men for their sins.[12] In the seventeenth and eighteenth centuries, the growth of mining, its economic significance, and the increase in the number of miners is reflected in the number of physicians interested in diseases affecting miners. There are no statistics for morbidity and mortality in this period, but George Rosen wrote that between 1600 and 1800 more than 25 authors contributed to knowledge of diseases that affected miners.[13]

Some authors who concerned themselves with the dust hazards confronting miners in particular were

1. Martin Pansa, author of *Consilium Peripneumaniaceum,* 1614
2. Ursinus, 1652, who believed dust inhalation caused peripneumonia
3. Stockhausen, 1656, who wrote about bergsucht (miners' phthisis)
4. Diembroeck, 1683, who described findings in stonecutters' lungs
5. Loehness, 1690, who described the marked effects of miners' work: "The dust and stones fall upon their lung, the men get lung diseases and at last take consumption."[14]

Bernardino Ramazzini, who made major contributions and, in the opinion of many, was the founder of occupational medicine, wrote in *De Morbis Artificium,* published in 1700, on the subject of diseases of stonecutters.

> We must not underestimate the maladies that attack stonecutters, sculptors, quarrymen and other such workers. When they hew and cut marble underground or chisel it to make statues and other objects, they often breathe in the rough, sharp, jagged splinters that glance off; hence, they are

usually troubled with cough, and some contract asthmatic affections and become consumptive. Moreover, from the marble tufa and stone emanates a maladic vapor which evidently severely affects the brain. They say that stonecutters who work with touch-stone are so affected in the head and stomach by the foul odor that it constantly exhales that they are sometimes forced to vomit. When the bodies of such workers are dissected, the lungs have been found to be stuffed with small stones. Diembroeck gives an interesting account of several stonecutters who died of asthma; when he dissected the cadavers he found, he says, piles of sand in the lung, so much of it that in cutting with his knife through the pulmonary vesicles he felt as though he were cutting a body of sand. He says too that he was told by a master stonecutter that when he was chiselling stone a dust arose, so fine that it penetrated the ox-bladders hanging in the workshop; in fact, in the course of one year he found that a handful of this dust had accumulated inside the bladder. The man declared that the dust would gradually prove fatal to stonecutters who take no precautions.[15]

The body of knowledge based upon observations of the association between dust and the risk of disease in mining and other dusty occupations continued to grow in volume but remained severely limited by lack of knowledge of specific dust diseases and lack of statistics on morbidity and mortality of workers in dusty trades. Silicosis was still unnamed, inadequately defined, and misunderstood.

INDUSTRIALIZATION AND SILICOSIS

As industry and mining expanded in the nineteenth century, the incidence of disease caused by dust grew and not surprisingly received renewed attention from the medical community. In 1870 Visconte coined the term *silicosis* to denote the pathological condition of lungs resulting from the inhalation of dust.[16] Interest in pigmented and dusty lungs of miners led to observation and study, and new contributions were made to the study of dust diseases.

Early in the nineteenth century, physicians believed that pulmonary pigmentation resulted from two separate entities, either natural black matter or *melanosis,* a term for a diseased lung, characterized by a deposition of black matter.

Some physicians attributed the origin of the pigmentation to an external source. Others theorized that blood alterations and other organic disturbances caused the pigmentation. Physicians' attention to pulmonary pigmentation among coal miners led many to believe that a noxious agent of extraneous origin was inhaled into the lung. Pearson published findings on lungs of coal miners in 1813.[17] Gregory then followed in 1831,[18] and both

Thompsons (father and son in 1837 and 1839) differentiated between phthisis and tuberculosis phthisis.[19]

In 1831, Thackrah began a study of the effects of dusty trades upon British workmen's longevity.[20] His important contributions to the literature included the fact that not all dusts shorten the lives of exposed workers. For example, he noted that while sandstone workers generally died before age 40, there seemed to be no unusual evidence of lung disease in brick and limestone workers. Greenhow in 1860 and 1861 investigated potteries, metal trades, cutlery making, and tin, copper, coal, and lead mining.[21] Peacock established the existence of miners' disease as an entity and distinguished it clinically from pulmonary tuberculosis.[22] In the *Transcriptions of the Pathological Society of London* (1860–1866), Peacock and Greenhow published clinical and pathological descriptions of the disease that Visconte later called *silicosis*.

Thus, by the end of the nineteenth century, the effects of dust inhalation on the lung had been observed and reported, and the significance of occupation as an influence on the occurrence of silicosis established. The manner in which silica dust exerted its effect upon the lung remained a matter of conjecture.

In the twentieth century, rapid industrialization caused economic, social, and technological change. Dangerous conditions in dusty trades, which had always existed, grew with the increase in production, new machines, and the larger numbers of workers in mines, factories, quarries, and other dusty industries. New technology—for example, the use of machines in mines—increased production. Mining machinery also increased the amount of dust. As the dangerous and unhealthy conditions increased with productivity and the growing number of exposed workers, the problem grew more pervasive and created a new interest in occupational diseases, particularly silicosis.

In the early decades of the twentieth century, Germans, British, and South Africans added to the knowledge of the effects of dust on lungs and the disease silicosis. In England in 1902, a governmental commission studied the high mortality among tin miners of Cornwall.[23] Haldane, a physiologist and member of the commission, attributed the illness of the miners to rock dust, which he said injured their lungs. In the same period in England, commissions and factory inspectors intensively investigated the refractories industry, the sandstone industry, potteries, metal grinding, and coal mining, all dusty industries with a high incidence of silicosis. Investigations pointed to free silica as the inciting agent in silicosis. Silicosis was also believed to be a factor that increased people's susceptibility to tuberculosis. As early as 1918, English workers received disability compensation for silicosis and tuberculosis.[24]

The research of the Miners' Phthisis Prevention Committee of South Africa found that the essential factor contributing to the development of chronic lung disease among gold miners was inhalation, over long periods, of dust containing free silica, generated in mining operations. The committee found that the majority of cases of early and intermediate silicosis were free from tuberculosis complications and that cases of silicosis can reach advanced stages and end in death without tuberculosis. They also found that all mineral dusts were not equally dangerous; the most dangerous dusts were characterized by the presence of uncombined crystalline silica.

It was discovered that silica, in order to cause disease, was in a state of minute subdivision as very fine, sharp-edged, insoluble particles. Seventy percent of silica particles in the lung were found to be less then one micron in diameter. In other words, only quartz dust generated the specific disease silicosis, and the smallest particles were the dangerous ones.[25] The South African research added further information to the technical picture of silicosis. In 1912, compensation for silicosis was first introduced in South Africa.

Little interest in silicosis existed in the United States until the second decade of the twentieth century. The newly aroused concern was coincident with the progressive movement, which sought protective labor legislation and workmen's compensation laws. A series of studies of mining and other dusty occupations revealed that silicosis was a severe health problem in the United States. The first of these studies, an investigation of the Joplin district in Missouri in 1914 and 1915, was made jointly by the United States Bureau of Mines and the United States Public Health Service and conducted by Lanza and Higgins.[26] Similar studies followed for other industries. In 1918 the granite industry received attention from the Committee on Mortality from Tuberculosis in the Dusty Trades, led by Frederick Hoffman.[27] In 1928 the Public Health Service conducted a study of the cement industry.[28] The sandstone, marble, pottery, and abrasives industries also received attention. In 1933 the United Mine Workers of America and the Pennsylvania Department of Labor and Industry surveyed pulmonary disease among anthracite miners.[29] The study confirmed what many had suspected: Pulmonary fibrosis among miners was due to silica dust, although coal might modify the clinical and pathological picture. Gardner and Cummings of the Saranac Laboratory studied pulmonary disease in the iron ranges of Michigan and Wisconsin.[30] Their studies indicated that the incidence of silicosis was related to the severity of exposure to quartz dust.

Laboratory research also extended knowledge of silicosis, and the use of roentgenological diagnosis of silicosis was applied. Engineering and chemical methods for dust determination and control were introduced. In

1922, Greenburg and Smith introduced the impinger, a dust-sampling device. The impinger method became the standard dust-sampling method used by the Bureau of Mines and the Public Health Service.[31] Drinker and Hatch and others also contributed to engineering methods of dust control.[32]

As a result of research and field studies, much was known about silicosis by the 1930s. For example, silicosis had been identified as an industrial disease resulting from the inhalation of silica dust. Its development depended upon the amount of free silica in the dust, the concentration of dust in a state of fine division in the air breathed, and the duration of exposure to the dust. Physicians had clinically described the disease and diagnosed it with x-rays, history of the patient's exposure, nature of exposure symptoms, and physical examinations. Prevention and control of silicosis by using engineering devices were understood. Dust-sampling methods could determine the amount of silica, leading to judgments of the severity of hazard and the adequacy of dust control.

Silicosis had become a preventable but uncontrolled occupational disease. Scientific knowledge had not yet been translated into a public policy of control. Between 1914 and 1930, compensation for occupational disease remained minimal. Up to the year 1935 not one state that operated under a schedule included silicosis in its list of compensable occupational diseases; in states with "all-inclusive" coverage for occupational diseases, the number of awards for silicosis was negligible.[33]

GAULEY BRIDGE

During the Great Depression, as the economic crisis deepened and the number of unemployed grew larger, workers were willing to face any known or unknown danger in order to receive a paycheck. It is not surprising that during these years industry felt little pressure to create hazard-free working conditions. Perhaps the most vivid and tragic illustration of the social situation is the Gauley Bridge incident.

In 1930 and 1931 the New Kanawha Power Company, a subsidiary of Union Carbide, contracted to drill a tunnel through a mountain at Gauley Bridge, West Virginia, to divert water from the New River to a hydroelectric plant. The subcontractor was Rhinehart-Dennis Company. The men employed to drill through the rock inhaled dust and soon complained of shortness of breath and other symptoms of silicosis. According to testimony, the employers knew the rock had a high silica content. The Rhinehart-Dennis Company employed an estimated 2,000 men from Pennsylvania, Georgia, North Carolina, South Carolina, Florida, Kentucky, Alabama, Ohio, and West Virginia (many of them black and unskilled).

Over a period of two years; 476 men died and eventually 1,500 were disabled.[34] The men received low pay; as the work progressed, the pay dropped even lower. Employees lived in squalid houses and worked under horrendous conditions. Descriptions of the dust are startling.

> The dust was so thick in the tunnel that the atmosphere resembled a patch of dense fog. It was estimated on the witness stand in the courtroom in Fayetteville, where suits against the builders of the tunnel were tried, that workmen in the tunnel could see only 10 to 15 feet ahead of them at times. Man after man—drillers, drill helpers, nippers, muckers, dinkey runners, and members of the surveying crew, who were the plaintiff's witnesses told of the dusty conditions. They said that although the tunnel was thoroughly lighted, the dinkey engine ran into cars on the track because the brakeman and dinkey runner could not see them. Laird King drove his dinkey into the little one and wrecked it, and Otis Edna, his brakeman, jumped off the front end just in time to save his life. Nippers who took charge of the steel bits could not see the sign by drillers when they needed "steel" and the signals had to be relayed. Dust got into the men's hair, on their faces, in their eyebrows; their clothing was thick with it. Raymond Jackson described how men blew dust off themselves with compressed air in the tunnel; if they did not they came out of the tunnel white, he said. One worker told how dust settled on the top of the drinking water.[35]

In 1936, after a subcommittee of the House of Representatives held hearings on the Gauley Bridge disaster it concluded that:

1. There was an utter disregard for all and any of the approved methods of prevention.
2. Dust was allowed to collect in such quantities that it lowered the visibility of the men.
3. The majority of drills were dry.
4. No appliances were used on drills to prevent concentration of dust.

The subcommittee stated, "The whole driving of the tunnel was begun, continued, and completed with grave and inhuman disregard of all consideration for the health, lives and future of the employees" and that "such negligence was either willful or the result of inexcusable and indefensible ignorance." The subcommittee also found that silicosis was prevalent in many states where mine and tunnel operations occur and "that silicosis is one of the greatest menaces among occupational diseases and that state laws governing prevention and compensation are totally inadequate."[36] The subcommittee recommended that Congress fund a study of silicosis, but to no avail. Secretary of Labor Frances Perkins convened a series of conferences on silicosis, but they accomplished very little. Punitive action was not taken against Rhinehart-Dennis, New Kanawha Power, or Union Carbide. Indeed, the hearings, conferences, and publicity did not lead to any effective action to curtail silicosis. In the case of Gauley Bridge, the failure to assume responsibility to protect the tunnel workers by applying established control measures caused workers to pay an inordinately high price.

World War II and the New Deal brought about new interest in occupational health. Postwar efforts to provide more salubrious working conditions and to alleviate the dust hazard took the form of continued investigations and research into the problem of silicosis, extension of workmen's compensation to cover occupational diseases in many states, and in some states the continuation of existing occupational health programs.

SILICOSIS AND DUST: PREVALENCE OF SILICOSIS

In 1950 the United States Department of the Interior, Bureau of Mines reviewed the literature on dusts, with emphasis on the relationship to dust diseases.[37] Efforts to control dust in industry employed the earlier conceived essential principle that there is a systematic dose-response relationship between severity of exposure to the dust hazard and the degree of response in the individual exposed. As the level of exposure decreases, there is a decrease in the risk of injury. Thus, the risk becomes negligible when exposure falls below certain tolerable levels. This concept assumes that harmful agents such as silica dust can be dealt with and that human exposure to this potentially harmful agent can be kept within tolerable limits. Control could then be effected by setting standards for acceptable levels of exposure, referred to by the American Conference of Governmental Industrial Hygienists (ACGIH) as Threshold Limit Values (TLVs).

The method of dust sampling and assessment in the United States from the 1920s to the 1960s was by the Greenberg-Smith or Midget Impinger. The impinger was used, followed by microscopic analysis, to obtain the particle count of airborne particles judged to be hygienically significant. A TLV adopted by the ACGIH was expressed as millions of particles per cubic feet (mppcf); it served as a guideline for compliance during these years. As the percent quartz in the dust increased, the allowable mppcf decreased according to a formula. In the late 1960s new dust assessment methods based on respirable dust mass were adopted. The permissible respirable mass of dust containing free silica is based on a formula that permits higher respirable masses as the percent of free silica in the dust is lowered. There are now standards for airborne dusts containing free crystalline silica based on respirable mass and/or particle count in OSHA and Mine Safety and Health Administration standards. There is little doubt that over the years the trend has been to lower permissible dustiness for dusts containing free silica.[38] The TLV for silica is currently highly controversial because there is evidence suggesting the absence of a threshold for silica and other hazardous substances.

The continuing studies of the environment of the granite quarries and sheds of Vermont and of the prevalence of silicosis among workers in the granite industry furthered understanding of the correlation between granite dust exposure and its effects. Over the years the incidence of silicosis in the granite industry was reduced as dust controls were introduced and dust concentrations drastically reduced. In 1925 Russell and colleagues indicated that among granite workers the first case of silicosis appeared after approximately two years of exposure to the highest average concentrations (59 mppcf). He found within four years a 100 percent prevalence of the disease and concluded that a safe level of exposure in the granite sheds was between 9 and 20 mppcf.[39] In 1937 Russell restudied the granite industry and again established the relationship between granite dust concentration and the progression of silicosis.[40] His data suggested that with an average dust exposure of 6 mppcf, unfavorable health effects to the worker would not occur. He suggested a Tentative Threshold of 10 mppcf. Another study of the granite industry in 1955 and 1956 by Hosey and others showed that engineering controls initiated in 1937, after Russell's study, had reduced granite shed dust concentration to about 5 mppcf and that airborne dust samples contained approximately 25 percent quartz.[41] Only one new, but doubtful, case of silicosis was found among men whose exposures began in 1937 or later. Workers employed prior to 1937 who were positive for silicosis averaged 32.4 years of exposure to granite dust; those without silicosis averaged 26.3 years of exposure. Of all the workers studied, half had started work before the initiation of dust controls. A longer time span needed to elapse before a final judgment could be made as to the efficacy of control measures and reduced dust exposure in relation to the prevalence and progression of silicosis.

In 1964 Ashe and Bergstrom further evaluated the dust controls instituted in 1937.[42] They did not find cases of silicosis in men whose exposure began in 1937 or later. Dust concentrations averaged 3 mppcf. In all the studies mentioned thus far, investigators used roentgenographic evaluations to determine the existence of silicosis in exposed workers.

Theriault and others carried out more recent studies of the Vermont granite industries (1969–1970). They based their findings on environmental data determined by personal respirable mass samples and both pulmonary function and roentgenographic evaluations. The studies developed a correlation of exposure and effect at relatively low average granite dust shed concentrations (523 μg per cubic meter for granite dust and 50 μg per cubic meter for quartz dust).[43-45] The studies suggested that early detection of dust effects in groups of workers would be better accomplished by pulmonary function tests than by roentgenographic evaluation. Effects were recorded at dust concentrations below the dust exposure standard of 10 mppcf used since 1937.

At the 1955 McIntyre-Saranac Conference on Occupational Chest Diseases, the Occupational Health Program of the Public Health Service presented some preliminary statistics indicating that silicosis still remained a widespread occupational health problem in the United States. A study covering the five-year period 1950 to 1954 disclosed that 10,362 cases of silicosis had been compensated or reported in 22 states. Of the 10,362 silicotics on record, 20 percent were dead, 50 percent were totally disabled, and 30 percent were still working, seeking work, or of an unknown status.[46] In indicating prevalence, not all states or even all potential sources of information with the 22 states were used. Thus, the study did not accurately reflect the number of persons with some form of silicosis.

> Despite the paucity and limitations of data one cannot deny that 10,362 cases is a sizeable number, even if spread over a five year period. Since we have no previous prevalence figures to relate it to or even an adequate estimate of the population risk to determine rates, we will have to be satisfied with this observation. This is a lot of silicosis for these times.[47]

Within the limitations imposed by the data, the findings supported the opinion that most cases of silicosis coming to light were among older men and represented a residue of old cases. The data also showed that silicosis was occurring among young men with recent exposure and pointed out the possibility that with less severe exposure silicosis took a longer time to develop. The study referred to silicosis as a serious "social and economic problem" because during the period 1950 to 1954 the number of compensation claims for silicosis increased. "Liberalization of compensation laws and improved diagnosis have undoubtedly influenced the situation. For instance, in West Virginia the number of claims filed increased from 447 in 1951 to 931 in 1954; in New Jersey from 18 in 1950 to 135 in 1954; in Alabama from 117 in fiscal 1952 to 298 in fiscal 1954."[48]

The Division of Occupational Health, the Public Health Service, and the Bureau of Mines reevaluated the prevalence of silicosis in the metal mining industry of the United States in 1958 to 1963. The project was an outgrowth of hearings before the Committee on Education and Labor of the House of Representatives in December 1956. The new study was the most extensive study made thus far in the metal mining industry in the United States. It was designed to determine the prevalence of silicosis among the work force of the metal mining industry. The following conclusions were reached:

1) Considerable progress has been made in the metal mining industry in the prevention of silicosis. The present overall prevalence rate of 3.4 percent is substantially lower than the rates encountered in previous surveys. The range varied among individual mines, however, from 0 to 12.9 percent and shows the prevalence of silicosis is not uniformly distributed throughout the industry. 2) The industry has instituted or improved many dust monitoring or dust control systems during the past twenty-five years; this has resulted in marked reduction in dust exposures. 3) There was a substantially higher prevalence of silicosis among men who worked in mining at some time before 1935 as compared with men who have worked the same number of years only since 1935. Two hundred and ninety-eight cases of silicosis were found in workers whose work history began prior to 1935. 4) One hundred and twenty-eight cases of silicosis were found among men exposed only since 1935. This finding, together with the excessive dust exposures in some of the mines studied, provides evidence that effective dust control has not been universally practiced. 5) Data obtained in the study do not permit judgment of the adequacy of present standards. In most of the mines studied, environmental data to define working conditions from 1935 to the time of this study were lacking. 6) Combined medical and environmental surveillance and control can prevent the development of clinically significant silicosis among miners.[49]

The 1939 study of non-ferrous metal mine workers in Utah was contrasted with the data in 12 lead-zinc mines of the 1958–1961 survey. Each had similar characteristics. Table 6-4 shows prevalence of silicosis in the western lead-zinc mine workers examined in 1958 to 1961 compared with Utah metal mine workers examined in 1939 according to the number of years worked in the metal mines. The table indicates a favorable trend in that the prevalence of silicosis for lead-zinc miners in the more recent survey was approximately 40 percent lower than that found in the 1939 Utah survey.

The prevalence data on silicosis in the later survey in most cases reflected exposures during the previous 20 or more years. The authors of the survey pointed out that

Without carefully collected and recorded environmental data on a periodic basis over the period of exposure, it is not possible with validity to assign weighted levels of dust exposure to workers. Such unfortunately is the case

TABLE 6-4.   Silicosis in Western Lead-Zinc Mine Workers and Utah Metal Mine Workers

Years at Metal Mines	Number Examined	All Silicosis		Simple Silicosis		Complicated Silicosis	
		Number	Percent	Number	Percent	Number	Percent
1939 Study							
Total	727	66	9.1	52	7.2	14	1.9
Less than 10 years	394	4	1.0	4	1.0	0	0
10–19 Years	228	30	13.2	26	11.4	4	1.8
20 Years and over	105	32	30.5	22	21.0	10	9.5
1958–61 Study							
Total	2,173	117	5.4	74	3.4	43	2.0
Less than 10 years	959	2	0.2	2	0.2	—	—
10–19 Years	717	26	3.6	17	2.4	9	1.2
20 Years and over	497	89	17.9	55	11.1	34	6.8

SOURCE: Reference 52.

for the 1958–61 survey. Only in a very few cases do the environmental findings of the survey apply retrospectively more than a few years. It will be another 10–20 years before the full impact of the environmental levels of dust exposure found in the 1958–61 survey reveal themselves in correlative silicosis prevalence data.[50]

OSHA AND AFTER

Without adequate quantitative description of the incidence of silicosis in America, it was impossible to assess and control the hazard. Once again in 1970, Americans learned that accurate statistics reflecting the incidence of silicosis did not exist and that no one knew the full extent of the silicosis problem. Testimony at hearings before the Subcommittee on Labor and Public Welfare of the United States Senate on the Occupational Safety and Health Act of 1970 described the shortcomings of the industrial health efforts in general and showed the need for a comprehensive federal occupational health and safety program. While acknowledging that efforts toward a hazard-free workplace had been made and some achievements had been realized, witnesses also stressed that there was still much to do. They sought provisions for research, collection of needed statistics, a method of recording and accumulating information, and the establishment and enforcement of meaningful standards.

Twenty years after the Occupational Safety and Health Act became law, exposure to silica dust and the risk of silicosis continues for many American workers. OSHA still has not promulgated a permanent standard for occupational exposure to crystalline silica that includes medical surveillance, record keeping, and control intervention requirements.

There are still some unanswered questions, including the extent of silicosis in the United States. We still do not know some of the fundamental aspects of silicosis. For example, what is the mechanism of the toxic action of silica? Dust suppression methods based on existing knowledge may be inadequate for control in all affected industries. We may need new and more accurate methods of collecting and measuring dust samples. However, even though we have unanswered questions, there is enough information to succeed in controlling silicosis through technology.

The National Institute for Occupational Safety and Health (NIOSH) recently estimated worker exposure to silica. Their survey indicated 147,569 plants with a total of 2,926,892 employees exposed to silica.[52] Table 6-5, based on NIOSH information, indicates deaths from silicosis for selected years between 1970 and 1986.

Unpublished data compiled from the Mine Health Research Advisory Committee in Tables 6-6 and 6-7 indicate mean age of death with silicosis,

TABLE 6-5.   Deaths from Silicosis for Males According to Age and Selected Years

Year	Number of Deaths											
	1970	1975	1978	1979	1980	1981	1982	1983	1984	1985	1986
25 Years and over	351	243	162	220	202	165	176	149	160	138	135
25–64	90	64	50	51	49	44	42	37	34	30	22
65 and over	261	179	112	169	153	121	134	112	126	108	113

SOURCE: *Health United States, 1988.* Publication (PHS) 89-1232. U.S. Department of Health and Human Services, Public Health Service, Centers for Disease Control. National Center for Health Statistics.

TABLE 6-6.   Deaths for White Males Attributed to Silicosis (1968–1985)

Mean Age of Death with Silicosis	Age 65 or Less	States Accounting for 61% of Silicosis-Related Deaths
72.9	16%	California Colorado Kentucky Michigan New Jersey New York Ohio Pennsylvania* West Virginia Wisconsin

* Pennsylvania had the most deaths.
Construction the most frequently listed industry.
Mining machine operator the most frequently listed occupation.

SOURCE: Unpublished data. Mine Health Research Advisory Committee Meeting (Summary of work in progress on silicosis surveillance) 1989.

TABLE 6-7.   Compensation Data for Silicosis (941 Reports of Silicosis in 27 States Between 1979 and 1983)

Males (%)	Over 55 (%)	Mean Age	States Reporting 94% of Cases	Industries with Most Reports	Other Industries with High Rates

90	70	55	Kentucky*
			Colorado
			Ohio
			New York
			Wisconsin
			North Carolina
				Bituminous coal
				(161 cases)
				Gray foundries
				(71 cases)
				Steel foundries
				(56 cases)
					Talc
					Soapstone
					Pyrophyllite
					    mining
					Clay
					    refractories
					Ferro-alloy
					    mining
					Lead and zinc
					    mining
					Industrial sand

* Kentucky had most cases (260 or 28% of total)
Mean annual incidence for mining was at least double that of nonmining.
Based on 941 reports of silicosis in 27 states, 1979–1983.

SOURCE: Unpublished data. Mine Health Research Advisory Committee Meeting (Summary of work in progress on silicosis surveillance) 1989.

states with highest percent of silicosis related deaths, most frequently listed industry and most frequently listed occupation (Table 6-6) in 1968 to 1985 and compensation data (Table 6-7) from 1979 to 1983. Silicosis still remains an industrial disease entity in the United States.

SUMMARY

The significance of the history of silicosis up to this time is that it highlights the present dichotomy between possession of sufficient technical knowledge to prevent the disease and the failure to eradicate it. In the past, social, economic, and political aspects of worker protection received little attention. The risks we once judged to be acceptable are now being reevaluated in the light of changed social attitudes.

Public policy toward occupational diseases such as silicosis now centers on issues of priorities and the basis for setting standards. Until there is scientific and political consensus, promulgation of a legally enforceable permanent standard will be highly controversial, enforcement will be difficult, and silicosis probably will not be eliminated from the roster of United States occupational diseases.

References

1. National Institute for Occupational Safety and Health. *Criteria for a Recommended Standard . . . Occupational Exposure to Crystalline Silica,* Publication 75-120. Washington: U.S. Department of Health, Education and Welfare, U.S. Public Health Service, Centers for Disease Control, National Institute for Occupational Safety and Health, 1974.
2. United States Department of Labor. *An Interim Report to Congress on Occupational Diseases.* Washington: U.S. Department of Labor, 1980.
3. World Health Organization Twenty-ninth World Health Assembly. Occupational Health Programme, Report by the Director General. April 1976.
4. Thrush, P. W., ed. *A Dictionary of Mining, Mineral and Related Terms.* Washington: U.S. Department of the Interior, 1968.
5. Hunter, D. *The Diseases of Occupations.* Boston: Little Brown & Co., 1969.
6. Stellman, J. M. and S. M. Daum. *Work Is Dangerous to Your Health.* New York: Vintage Books, 1973.
7. *Encyclopaedia Britannica,* 15th ed., 1977.
8. International Agency for Research on Cancer. World Health Organization. *Evaluation of the Carcinogenic Risk of Chemicals to Humans: Silica and Some Silicates,* Vol. 42 Lyons, France: World Health Organization, 1987.
9. Lanza, A. J., ed. *Silicosis and Asbestosis.* New York: Oxford University Press, 1938.
10. Rosen, G. *The History of Miners' Diseases.* New York: Schumans, 1943.
11. Agricola, G. *De Re Metallica.* Trans. H. C. Hoover and L. H. Hoover New York: Dover, 1950.

12. *Encyclopaedia Britannica*, 15th ed., Vol. 13.
13. Rosen, *History of Miners' Diseases*, 32.
14. Teleky, L. *History of Factory and Mine Hygiene*. New York: Columbia University Press, 1948.
15. Ramazzini, Bernadino. *Diseases of Workers*. New York: Hefner Publishing Co., 1964.
16. Hunter, *Diseases of Occupations*, 958.
17. Pearson, G. "On the Colouring Matter of the Black Bronchial Glands and of the Black Spots of the Lungs." *Philosophical Transactions of the Royal Society* 103(1813): 159.
18. Gregory, J. C. "Case of Peculiar Black Infiltration of the Whole Lungs, Resembling Melanosis." *Edinburgh Medical and Surgical Journal* 36(1831): 389.
19. Thompson, W. "On Black Expectoration and the Deposition of Black Matter on the Lungs." *Medical-Chirurigal Transactions* 20(1837): 230–31.
20. Thackrah, C. T. *The Effects of the Principal Arts, Trades and Professions, and of Civic States and Habits of Living on Health and Longevity*. London: Longmans, 1831.
21. Meiklejon, A. "History of Lung Diseases of Coal Miners in Great Britain: Part I, 1800–1875." *British Journal of Industrial Medicine 8(1951): 127–137.*
22. Lanza, A. J. *Silicosis and Asbestosis*, 5.
23. Haldane, J. B. S., J. S. Marint, and R. A. Thomas. *Report to the Secretary of State for the Home Department on the Health of Cornish Miners*. London: His Majesty's Stationery Office, 1904.
24. Air Hygiene Foundation of America. *Silicosis and Allied Disorders*. Medical Bulletin 1. Pittsburgh: Air Hygiene Foundation of America, 1937.
25. Teleky, *History of Factory and Mine Hygiene*, 203.
26. Higgins, E., A. Lanza, F. B. Laney, and G. S. Rice. *Siliceous Dust in Relation to Pulmonary Disease among Miners in the Joplin District, Missouri*. Bulletin 132. Washington: U.S. Bureau of Mines, 1917.
27. Hoffman, Frederick L. *Mortality from Respiratory Diseases in Dusty Trades*. Bulletin 231 Washington: U.S. Bureau of Labor Statistics, 1918.
28. Thompson, L. R., D. K. Brundage, A. E. Russell, and J. J. Bloomfield. *The Health of Workers in Dusty Trades, 1. Health of Workers in a Portland Cement Plant*. U.S. Public Health Service Bulletin 176. Washington: U.S. Public Health Service, 1928.
29. Sayers, R. R., J. J. Bloomfield, J. M. DallaValle, R. R. Jones, W. E. Dressen, D. K. Brundage, and R. H. Britten. *Anthraco-Silicosis Among Hard Coal Miners*. U.S. Public Health Service Bulletin 221 Washington: U.S. Public Health Service, 1935.
30. Lanza, *Silicosis and Asbestosis*, 20.
31. Ibid., 21.
32. Drinker, P. and T. Hatch. *Industrial Dust*. New York: McGraw-Hill, 1954.
33. Lanza, *Silicosis and Asbestosis*, 409.
34. Hearings Before a Subcommittee of the Committee on Labor, House of Representatives. *H.J. Res. 449*, 74th Congress, 16, 17, 20, 21, 27, 28, and 29

January; 4 February 1936. Washington: U.S. Government Printing Office, 1936.

35. Ibid., 4.

36. Ibid., 201–2.

37. Forbes, J. J., S. J. Davenport, and G. Morgis Genevieve. *Review of Literature on Dusts*. U.S. Department of Interior, Bureau of Mines, Bulletin 48. Washington: U.S. Government Printing Office, 1950.

38. American Conference of Governmental Industrial Hygienists. *Documentation of the Threshold Limit Values for Substances in Workroom Air*. 3d ed. Cincinnati: ACGIH, 1971.

39. Russell, A. E., R. H. Britten, L. R. Thompson, and J. J. Bloomfield. *The Health of Workers in Dusty Trades: II, Exposure to Silicosis Dust (Granite Industry)*. Public Health Bulletin 187 Washington: U.S. Treasury Department, Public Health Services, 1929.

40. Russell, A. E. *The Health of Workers in Dusty Trades: VI, Restudy of a Group of Granite Workers*. Public Health Bulletin 269. Washington: Federal Security Agency, Public Health Service, 1941.

41. Hosey, A. D., H. B. Ashe, and V. M. Trasko. *Control of Silicosis in Vermont Granite Industry—Progress Report*. Public Health Publication 557, Washington: U.S. Department of Health, Education and Welfare, Public Health Service, 1957.

42. Ashe, H. B. and D. E. Bergstrom. "Twenty-six Years Experience with Dust Control in the Vermont Granite Industry." *Industrial Medicine and Surgery* 33(1964): 973–978.

43. Theriault, G. P., W. A. Burgess, L. J. DiBerardinis, and J. H. Peter. "Dust Exposure in the Vermont Granite Sheds." *Archives at Environmental Health* 28(1974): 12–17.

44. Theriault, G. P., J. M. Peters, and L. J. Fine. "Pulmonary Function in Granite Shed Workers of Vermont." *Archives of Environmental Health* 28(1974): 18–22.

45. Theriault, G. P., J. M. Peters, and W. H. Johnson. "Pulmonary Function and Roentgenographic Changes in Granite Dust Exposure." *Archives of Environmental Health* 28(1974): 23–27.

46. Trasko, V. H. "Some Facts on the Prevalence of Silicosis in the United States." *American Medical Association Archives of Industrial Health* 14(1956): 379.

47. Ibid., 382.

48. Ibid., 386.

49. Flinn, R. H., et al. *Silicosis in the Metal Mining Industry, A Reevaluation 1958–1961*. Public Health Service Publication 1076 Washington: U.S. Government Printing Office, 1963.

50. Ibid., 168.

51. Ibid., 167.

52. National Institute for Occupational Safety and Health. unpublished material. National Occupational Exposure Survey (NOES), 23 February 1990.

Chapter 7

Vinyl Chloride

The chemical industry covers an enormous span of products and technologies, from the manufacture of pharmaceutically valuable materials in small quantities to bulk chemicals utilized in the production of plastics. These ubiquitous chemicals provide enormous benefits. At the same time, they introduce risks to the health and well-being of workers and the general public. Increased exposure of the population to chemical hazards occurs during manufacturing in the occupational environment, in the outdoor environment where effluents to air, water, or solid wastes are encountered by the public, and at the commercial product user stage. The magnitude of the growth of chemical usage and awareness of the associated inherent potential dangers have raised a number of complex and controversial questions, brought the debate on toxic chemicals into the public arena, and made chemical pollution a matter of public health policy.

Vinyl chloride, an important component of this significant industry, represents one of the many chemicals once considered safe but now known to cause cancer and other chronic illnesses. Because of the pervasive nature of vinyl chloride, five major federal agencies operating under approximately 15 separate health and environmental statutes regulate that chemical.[1]

This chapter focuses on one of the many health policy decisions related to regulation of chemicals in the United States, namely, setting a workplace standard for control of occupational exposure to vinyl chloride in 1974 by the Occupational Safety and Health Administration, the federal agency charged with control of occupational hazards. It is a case study designed to illustrate the interplay between scientific and policy determinations during the regulatory process.

SCIENCE AND POLICY

Events before, during, and after the vinyl chloride hearings, or rule-making proceedings, demonstrate the intricate relationship between science and policy as well as the collision of opposing interests that accompany and complicate decision making in the regulatory arena. The roots of contention over regulation of vinyl chloride are both scientific and political. They can be found in changes of public expectation and social values, new scientific ability to make more accurate measurements, increased utilization of chemicals, and the uncertainties and complexities involved in identifying and assessing work-related disease.

In 1974 the issue—whether to regulate vinyl chloride, an undisputed carcinogen—was resolved. At the time the controversy really concerned how to assess the risk associated with vinyl chloride and what degree of control to seek. The scientific problems included the determination of what constituted a precise level of exposure, whether a safe level of exposure existed, and whether animal data could be extrapolated to humans in quantitative terms.

The policy issue included determination as to the appropriate margin of safety, tolerable level of risk, and how certain one must be about the data before considering the extent of regulation. The interplay between policy and science was most apparent in the need for scientific data on which to base sound policy judgments of how, when, and why a substance is to be regulated.

The task of developing a workplace health standard was further complicated because deliberations occurred under intense pressure from divergent groups. The Occupational Safety and Health Administration utilized public proceedings, i.e., rule-making proceedings, to provide scientific review. The process brought science into the public arena and stimulated a controversy that focused on the questions of cost and technological feasibility. The data were often incomplete, presented in an adversarial atmosphere, and susceptible to a variety of interpretations brought about by the conflicting interests and values of participants at the hearings. Indeed, the policy issues debated indicated a deep philosophical controversy over what is just and fair and whether one can or should balance the cost of controls against the risks of ill health.

All this occurred in an atmosphere pervaded by crisis, precipitated by the discovery of an excess incidence of a rare cancer, angiosarcoma of the liver, in vinyl chloride workers. The characteristic trait of chemical carcinogenesis, existence of a latent period of 20 or more years between initial exposure and the onset of cancer, led to fear that large numbers of people had been exposed to vinyl chloride prior to affirmation of its carcinogenicity in 1974.

In the 1970s public concern about the effects of technologies on the environment was heightened as scientific evidence that chemical substances could induce cancer or other adverse chronic health effects in humans emerged. Regulatory agencies such as OSHA focused attention upon cancer. The process of assessing the risk of cancer and other adverse health effects from exposure to vinyl chloride raised a number of scientific and broad social policy questions, highlighted basic and still unresolved issues, and focused public attention upon the difficulties in decision making for the regulation of carcinogens. Coping with scientific uncertainties during the process of making public health policy decisions, along with the problem of how to balance economic and social needs, continues to create controversy during the decision-making process to regulate chemicals in the environment. Both issues were debated during the vinyl chloride hearings. That debate set the stage for future health policy decisions and revealed the interaction between politics and science.

In the mid-1970s Lowrance defined *safety* in the following manner: "A thing is safe if its attendant risks are judged to be acceptable."[2] Thus, determining safety involves two extremely different kinds of activity. The first, measuring risk, that is, the probability and severity of harm, is a scientific activity. It utilizes quantification and implies objectivity. The second activity, judging risk or judging the acceptability of risk, is a political activity that implies social value.

Since Lowrance defined and analyzed risk, the definitions of *risk* have been explored and refined in an endeavor to create a framework for decision making. A National Academy of Sciences study suggests that we should make a distinction between risk assessment and risk management. The study concluded that "the scientific findings and policy judgments embodied in risk assessment should be explicitly distinguished from the political, economic and technical considerations that influence the design and choice of regulatory strategies."[3]

This is a retrospective view of the events and issues that pervaded the decision-making process for the regulation of vinyl chloride in the workplace. The case study begins with background material on vinyl chloride production and its effects on health.

VINYL CHLORIDE PRODUCTION AND EFFECTS ON HEALTH

Commercial production of vinyl chloride began in the United States in the 1930s, although it was first synthesized in 1837. After World War II production levels rose rapidly, increasing from 45 million kilograms in 1943 to

2.4 billion kilograms in 1973.[4] At that time 17 plants employing approximately 940 workers produced vinyl chloride. Forty plants produced polyvinyl chloride (PVC) and employed approximately 5,600 workers.[5] The greatest potential for harmful exposures occurs when vinyl chloride is polymerized to form polyvinyl chloride resins. The principal use of vinyl chloride is in the production of polyvinyl chloride, which in turn is utilized to manufacture a wide variety of plastic materials including floor tiles, paints, phonograph records, pipes, electric insulation, furniture, upholstery, draperies, wall coverings, tablecloths, footwear, food packages, garden hoses, dentures, bottles, pharmaceutical products and numerous other consumer products.

Prior to 1974 knowledge of adverse effects associated with human exposure to vinyl chloride came from occupational observations. Human cancer was not associated with vinyl chloride before the 1970s, but other hazardous characteristics of the chemical were known. Its explosive and highly flammable nature could cause the apparent danger of explosion or fire. In the 1940s vinyl chloride gas was utilized for medical anesthesia but quickly abandoned because it caused arrhythmia of the heart. From 1960 through 1963 published scientific reports documented anesthetic effects in animals and humans and liver injuries in animals from chronic exposures.[6] The 1962 edition of Patty's *Industrial Hygiene and Toxicology* summed up the accepted and then known adverse effects of vinyl chloride.

Vinyl chloride appears to be a material of extremely low toxicity. The principal response seems to be one of central nervous system depression, which may result in symptoms of dizziness and disorientation that are somewhat similar to the response from ethyl chloride exposure. There is possibility of some lung irritation occurring from chronic exposure as some edema is observed in acute vapor exposure. Most investigators do not observe kidney or liver damage. One group of authors indicated some hyperemia of the liver and kidneys from acute exposure. It is concluded that the material has essentially a narcotic effect, with some lung irritations and a possibility of organ injury. There has been quite extensive use of this material in the chemical industry but no clinical reports of injury.[7]

In the 1960s a new clinical entity associated with vinyl chloride exposure appeared among workers engaged in the process of polymerization of vinyl chloride to polyvinyl chloride. The disease, acro-osteolysis, included symptoms of tenderness of fingertips, sometimes accompanied by gradual destruction of the bony integrity of the fingers. Workers with this condition exhibited a form of vascular disease known as Raynaud's phenomenon.[8] In the 1970s evidence of liver disease appeared among vinyl chloride workers.[9] Effects of exposures to vinyl chloride were lumped together and labeled *vinyl chloride disease*.[10] Viola and Maltoni first associated neoplasia with vinyl chloride in 1971. They later reported malignant changes in animals, including angiosarcoma.[11]

OSHA: U.S. REGULATORY APPROACHES TO WORKPLACE CHEMICALS

In order to understand what occurred in 1974, when confirmed cases of angiosarcoma appeared among vinyl chloride workers, it is necessary to examine the progress of OSHA in preventing diseases in the workplace and protecting workers. When Congress passed the Occupational Safety and Health Act in 1970 it gave OSHA in the Department of Labor the power to set and enforce mandatory occupational safety and health standards. The law also established the National Institute of Occupational Safety and Health (NIOSH) within the Department of Health, Education, and Welfare to gather and analyze scientific data for the formation of occupational safety and health standards. During the first four years of its existence, OSHA emphasized safety issues and paid little attention to health. Yet in 1972 NIOSH estimated that 390,000 new cases of occupational disease occurred each year.[12] This OSHA policy of safety first was reflected in the small number of health standards, the inadequate number of health inspections, and the lack of qualified health inspectors. OSHA also failed to implement sections of the act requiring that employees be apprised of and protected from hazards to which they are exposed and those sections requiring employers to maintain accurate records of employee exposures to potentially toxic materials and to supply employees with access to these records.[13]

By 1974 OSHA's total output of adopted standards added up to two: asbestos and a standard for 15 carcinogens. The General Accounting Office estimated that at the Labor Department's current rate of promulgating standards, it would take 100 years or more to develop needed health standards.

OSHA policy reflected administrative decision making. The agency, through the standard-setting procedure, had leeway to construct and carry out the technical details of policy enunciated by the Congress in the Occupational Safety and Health Act.

In 1971 OSHA had adopted national consensus standards and/or previously existing federal standards in order to guarantee a minimum level of protection until permanent standards could be enacted. The consensus standards covered permissible airborne exposure concentrations for approximately 400 substances that had exposure limits set in 1968 by the American Conference of Government Industrial Hygienists, subsequently incorporated in the Walsh-Healy Public Contracts Act for contractors doing business with the federal government. OSHA adopted these standards on the basis of preexisting federal law. The standards were considered limits of airborne concentrations for chemical substances that nearly all workers could be exposed to for 40 years, 50 weeks per year, without adverse effect. The estimated concentrations were called Threshold Limit Values (TLVs).[14] TLVs codified contemporary industrial practice and were set at concentrations necessary to prevent acute and chronic effects. The TLVs were attacked from two sides, those who thought them too permissive and others who thought them stricter than necessary. The list of 400 TLVs included vinyl chloride at 500 parts per million parts of air (ppm). Organized labor stressed the weaknesses and inability of national consensus standards to protect workers. They objected to lack of adequate forms of warning and failure to provide for employee access to employer monitoring records. OSHA had assured labor that those provisions would be added by the end of the summer of 1971.[15] They did not fulfill their promise.

The AFL-CIO maintained that OSHA's inactivity after adopting the TLV list could be explained as a deliberate action on the part of the Assistant Secretary of Labor for Occupational Safety and Health, brought about by improper political pressure. They cited the Guenther memo of June 14, 1972. This memo from Assistant Secretary Guenther implied that a properly managed OSHA could be used as a sales point for fundraising in the presidential election of 1972. The memo referred to the cotton dust standard and promised that OSHA would not propose highly controversial standards.[16] Labor referred to this memo as the "Watergate memo."[17] The AFL-CIO believed a good law was being poorly administered, and that rather than seeking administrative changes, congressional reaction was to destroy the law by appropriating inadequate funds.[18] Critics also noted that there was a systematic replacement of skilled career civil servants with unqualified political appointees, that OSHA delayed standards, and that progress in health and safety came only as a result of pressure from labor. Clearly OSHA was beset with problems. Too few health professionals, poor leadership, lack of administrative commitment, and political interference influenced the setting of standards and the enforcement powers of the agency.

DISCOVERY OF ANGIOSARCOMA AND RESULTING ACTIONS

Then, in January 1974, the B.F. Goodrich Company, a manufacturer of polyvinyl chloride, notified its employees, NIOSH, and the Kentucky

Department of Labor that three workers had died of angiosarcoma of the liver, a rare and incurable cancer. Because all three workers had employment in the manufacture of polyvinyl chloride resins as a common denominator, the suspicion arose that vinyl chloride caused angiosarcoma of the liver, and that it was occupationally related.[19]

Public disclosure caused widespread concern about the harmful effects of vinyl chloride and led to a flurry of activity. It became the first of a number of chemical carcinogen crises. Policy decisions on control of vinyl chloride in the workplace would be influenced by labor, industry, government, the press, public interest groups, and the scientific community.

After the announcement by B.F. Goodrich, OSHA and NIOSH conducted a walk-through survey of the Kentucky plant and verified the causes of death. Pressured by labor and public interest groups, OSHA called for a fact-finding hearing on February 15. The hearing included representatives from government, labor, industry, and the scientific community. Articles about vinyl chloride appeared in newspapers and national magazines. The Oil, Chemical and Atomic Workers, United Rubber Workers, United Steel Workers, the Industrial Union Department of the AFL-CIO, and public interest groups exerted pressure on OSHA to promulgate a vinyl chloride standard.

A number of factors account for the activity that had ensued. First, the public's perception of cancer as more serious than other diseases, irreversible once initiated and often fatal, caused fear and deep concern. Angiosarcoma of the liver was one of the rarest human malignant neoplasms with previous known incidence in the United States of only approximately 21 cases a year.[20] After B.F. Goodrich disclosed its cases of angiosarcoma, other companies also reported deaths from angiosarcoma among vinyl chloride and polyvinyl chloride workers.

Second, the growing number of known cases led to the fear that only the tip of the iceberg had appeared. The national cancer survey expected the annual incidence of this tumor to be about 25 to 30 cases a year in the entire United States population. By estimating the total vinyl chloride worker population, past and present, in the United States at roughly 20,000, it was calculated that about 0.03 cases of this tumor might be expected in such a population over a decade. Epidemiologists extrapolated from the number of cases discovered in 1974 and concluded the risk ratio (the ratio of observed to expected cases) was at least 400 : 1.[21] The tumors had appeared in the Louisville workers 14 to 27 years from the onset of their exposure. This long latent period meant that past uncontrolled exposures would be reflected in tumors in decades to follow.

The third factor, public awareness, forced OSHA into action, and finally pressure from labor forced OSHA to act. The Industrial Union

Department of the AFL-CIO had immediately contacted the unions representing vinyl chloride workers; prepared information for workers; consulted with experts from Mt. Sinai School of Medicine, University of Pittsburgh, Howard University, University of North Carolina, and Case Western Reserve; and initiated a nationwide study. The AFL-CIO was also represented at the meeting of NIOSH, industrial officials, and scientists.[22] It is clear that labor intended to make its stand for a more viable OSHA on the vinyl chloride issue.

In February NIOSH and Centers for Disease Control (CDC) briefed other federal agencies with health research responsibilities. NIOSH also began to develop a recommended occupational health standard for vinyl chloride.

On February 15, OSHA held an Informal Fact Finding Hearing on Possible Hazards of Vinyl Chloride Manufacture and Use, to determine whether the situation warranted an emergency standard. Industry wanted any further controls of exposure to vinyl chloride set by ordinary rulemaking procedures. The AFL-CIO argued that the confirmed cases of angiosarcoma warranted an Emergency Temporary Standard (ETS). This is significant because, by law, if OSHA issues a temporary standard, then legally a permanent one must follow in 6 months. Thus, industry supported improving the current federal (TLV) standard of 500 ppm. Their position was that rulemaking would allow for orderly development of data.

On April 5, 1974 OSHA promulgated an ETS that reduced the Permissible Exposure Level (PEL) from 500 ppm to 50 ppm.[23] The agency also established other temporary requirements such as workplace monitoring, use of air-supplied respirators if the new PEL is exceeded, and impervious suits in some instances. OSHA based PEL reduction on the findings of Cesare Maltoni, whose studies showed induction of angiosarcoma of the liver and other organs and the production of other cancers in rats exposed to vinyl chloride. His research confirmed vinyl chloride as a carcinogen and further confirmed its role in inducing the cancers observed in B.F. Goodrich workers.[24] NIOSH had recommended a temporary standard of no measurable concentrations of vinyl chloride. Apparently the 50 ppm emergency standard was a compromise between the wishes of labor and those of industry. OSHA justified its emergency standard on economic and medical grounds.

In response to the vinyl chloride crisis a workshop entitled "Toxicity of Vinyl Chloride–Polyvinyl Chloride" sponsored by the New York Academy of Sciences, the American Cancer Society, the National Institute of Environmental Health Sciences, NIOSH, and the Society of Occupational and Environmental Health convened a large international working group of scientists to disseminate information on the toxicity and carcinogenicity of vinyl chloride.[25] Scientists at the workshop presented a great deal of information linking vinyl chloride to angiosarcoma of the liver and to a variety of other diseases and tumors in sites other than the liver.

NIOSH began a major epidemiological study of vinyl chloride workers. The Manufacturing Chemists Association (MCA) also sponsored a study. Meanwhile, the number of confirmed deaths due to angiosarcoma among vinyl chloride workers grew.

On the basis of evidence at the February hearings, the demonstrated evidence of the carcinogenicity of vinyl chloride in animal studies, the epidemiological studies, and public pressure, OSHA proposed a permanent standard of "no detectable level" on May 18, 1974.[26] It required monitoring employee exposures, engineering and work practice controls, medical surveillance, protective clothing, emergency procedures, and training. OSHA set the hearing dates for June and July, allowing a period of 30 days for public comment.

HEARINGS FOR A PERMANENT STANDARD

Records of the eight days of hearings include prehearing and posthearing comments, testimony at the fact-finding and rule-making hearings, records of OSHA inspections, an environmental impact statement, and economic and technical impact studies. OSHA received more than 600 written comments and more than 200 oral and written submissions. The record exceeds 4,000 pages. Employers, employees, labor unions, public affairs groups, physicians, and scientists submitted information and testified at the long and acrimonious hearings. The adversarial relationship between labor and industry led each side to take opposing positions on the following key issues: Could the proposed exposure limit be supported on medical grounds? Was the proposed standard technologically and economically feasible?

No one questioned the carcinogenicity of vinyl chloride, but representatives of industry believed that evidence did not support the "no detectable" level or 1 ppm exposure ceiling proposed by OSHA. The Society of the Plastics Industry (SPI) maintained that data did not exist to prove the exposure provided under the ETS (50 ppm) unsafe. Spokesmen from industry also pointed out that conclusions about human sensitivity to vinyl chloride could not be drawn from animal tests. They suggested the existence of a threshold for human carcinogenesis.[27]

In contrast, cancer specialists from universities, the National Cancer Institute, and NIOSH and persons appearing for unions and public interest groups said that safe doses of carcinogens could not be scientifically determined.[28] Experts testified that quantification of a safe concentration was not possible with the present state of scientific knowledge.

Representatives from industry testified that the permanent standard was not technologically feasible and would shut down the plastic industry if adopted.[29] SPI estimated that a shutdown would result in the loss of 1.7 million to 2.2 million jobs and loss of production valued at $65 million to $90 million annually.[30] SPI recommended less stringent standards.

Labor unions and the Health Research Group suggested that levels below 1 ppm were attainable and that means existed, or could be found, to maintain exposure below 1 ppm. Labor accused industry of attempting to blackmail OSHA into setting a lenient standard by exaggerating the difficulty and the cost of compliance and by predicting economic disaster and dislocation.[31]

On October 4, 1974 OSHA promulgated the permanent standard for vinyl chloride. It accepted the principle of 1 ppm as the maximum possible exposure. OSHA based its conclusion upon thorough review and evaluation of evidence submitted. However, where factual certainties were lacking or where facts alone did not provide an answer, OSHA made policy judgments. The hearings did not determine the precise level of exposure that poses a hazard. It did not answer the question of whether a "safe level" exists; nor did the hearings clearly determine to what extent exposures could be feasibly reduced. OSHA made the policy decision not to wait for indisputable answers because lives of employees were at stake. OSHA exercised its "best judgment" on the basis of available evidence. The judgments required a balancing process in which OSHA's overriding consideration was the protection of employees.[32]

The SPI and other companies petitioned the federal courts for a stay of implementation of the standard. In *Society of the Plastics Industry, Inc.* v *OSHA* and *Firestone* v. *United States Department of Labor,* the Second Circuit Court reviewed regulations limiting worker exposure to vinyl chloride. The court stated,

> As in Industrial Union Department AFL-CIO v Hodgson supra, the ultimate facts herein disputed are on the frontiers of scientific knowledge, and though the factual finger points, it does not conclude. Under the command of OSHA, it remains the duty of the Secretary to act to protect the working man, and to act even in circumstances where existing methodology or research is deficient. The Secretary in extrapolating the M.C.A. study's findings from mouse to man, has chosen to reduce the permissible level to the lowest detectable one. We find no error in this respect.[33]

The petition was denied and the United States Supreme Court refused to review the decision.

AFTER THE STANDARD

Representatives of the plastics and chemical industries had argued that if OSHA issued a permanent standard for vinyl chloride of 1 ppm the entire polyvinyl chloride industry would shut down. Instead, some of the control technology developed to reduce worker exposures to within allowable limits also cut costs and increased productivity by reducing worker exposures, improving product quality, reducing the time for cleaning the reactor vessels, and combining previously separate processes. New production techniques were also sold to other companies. For example, B.F. Goodrich licensed its processes to other chemical manufacturers.[34] By the time the standard became effective, major companies admitted that they could operate without curtailing production. Threatened shutdowns did not occur, and major vinyl chloride producers licensed a variety of emission-control devices. Industry's forecast of curtailment of production and economic ruin did not occur.

By 1979 vinyl chloride production capacity had increased by 41 percent and polyvinyl chloride capacity by 85 percent over 1974 levels. Expansion of vinyl chloride and PVC produced since 1974 has created an estimated 2,000 jobs, offsetting the job losses estimated by spokesmen for the chemical and plastics industries. In the years immediately after the issuance of the standard, the growth rate for the vinyl chloride industry was above the average for United States industry, and profits increased.[35]

It is not possible to judge the long-term health effects of the vinyl chloride standard by using any measures of impact because of the long latent period of carcinogenic action. However, a discussion of effects can be based on surrogate measures of effectiveness. OSHA's accomplishments can be reviewed by utilizing as surrogate measures compliance efforts and reduction of exposure levels for employees exposed to carcinogens on the job.[36] In the case of vinyl chloride in 1973, exposure varied from 50 to 500 ppm. After OSHA promulgated its permanent standard for a PEL of 1 ppm in 1974, employees in the vinyl chloride industry could be considered to have experienced a subsequent exposure reduction of 98 percent. This reduction in exposure can be considered a measure of cancer reduction because the dose-response curves for carcinogens are assumed to be linear and to extend to zero in the absence of a threshold for the effect.

Thus, the facts proved to be contrary to private sector manufacturing projections. It proved possible to have health and safety standards that protect working people and, at the same time, to increase productivity.

The case of vinyl chloride provides serious erosion of the pessimistic position that health and safety standards are a deterrent to productivity or that they have a negative impact upon productivity. In the case of vinyl chloride, the private sector did not do its homework and miscalculated the effects of a significantly lowered standard on the industry. A repercussion of this miscalculation is that economic feasibility became one of the most difficult issues to resolve in future OSHA standards proceedings. During the standard-setting hearings for vinyl chloride, it was stated that lowering the standard for vinyl chloride might reduce cases of angiosarcoma but it would also close down the industry. That did not happen. The industry continued to grow in spite of the lower standard.

CONCLUSIONS

On the surface there is a lesson in the events of 1974. Vinyl chloride brought public attention to occupational health hazards and created a growing awareness of the influences of industrial processes upon health. Forces were mobilized to define, evaluate, and translate information into action to create a public health policy for control of a dangerous chemical. Concern was mobilized into action, and the necessary standards were promulgated to protect workers. The catalyst, public information, led to regulation of this pervasive chemical.

The more complex lesson is in the legacy of the vinyl chloride hearings. In 1974 the chance recognition of an occupational cancer, under a unique set of circumstances, pointed to the problems of carcinogenic chemicals in the workplace, demonstrated what was unknown about chemical carcino-genesis, and focused attention on the absence of a national policy for occupational cancer.

The hearings that ensued undoubtedly affected future efforts to regulate the impact of industrial chemicals on health. At the time OSHA did not have formal guidelines for risk assessment. They utilized rule-making procedures to provide scientific review as a substitute for independent peer review, which led to criticism of the agency and focused attention on policy and politics rather than on science. Since 1974 increasing pressure has been placed on regulators to justify proposed regulations. They are asked, What risks will be removed and reduced? What benefits will be gained? What will it cost? The unrealistic industrial estimates of the cost of controlling vinyl chloride and the unavailability of creditable cost figures in 1974 led to increased emphasis on cost effectiveness and cost-benefit analysis during subsequent standard-setting proceedings. Lack of infor-mation about levels of occupational exposures to vinyl chloride has led to increased surveillance of exposed populations, better record keeping, and analysis of records.

Two Supreme Court decisions, *Industrial Union Department AFL-CIO* v. *American Petroleum Institute* (1980) and *American Textile Manufac-*

turers Institute, Inc. v. *Donovan* (1981), would later address the unanswered issues that were part of the legacy of the vinyl chloride standard, a legacy that impacted on future efforts to regulate toxic substances in the workplace. The questions of costs and benefits, scientific uncertainties, and feasibility raised earlier at the vinyl chloride hearings would be decided by the Supreme Court in the benzene and cotton dust rulings.

When the Fifth Circuit Court invalidated the OSHA benzene standard, the court contended that OSHA had failed to provide a quantitative estimate of the benefits to be achieved by reducing the PEL from 10 to 1 ppm. The Fifth Circuit Court held that the Secretary must determine "whether the benefits expected from the standard bear a reasonable relationship to the costs imposed by the standard."[37] The case was sent to the Supreme Court. Although a divided Supreme Court issued a number of separate opinions, the majority affirmed the decision of the Fifth Circuit Court that there must be a reasonable relationship between cost and benefit. Furthermore, the majority said that "the burden was on the Secretary [agency] to show, on the basis of substantial evidence, that it is at least more likely than not that long-term exposure to 10 ppm of benzene presents a significant risk of material impairment."[38] The Supreme Court shifted the burden of proof to the agency. The need to prove significant risk when evidence of harmful effects is often inconclusive and subject to dispute precluded OSHA from adopting new standards unless definitive, detailed, and indisputable evidence existed. It did not solve the dilemma of what to do when evidence clearly shows that at some levels a substance causes serious illness.

In 1981 in *American Textile Manufacturers Institute, Inc.* v. *Donovan* (cotton dust standard), the Supreme Court once again addressed the question of whether the Secretary was required, when promulgating a standard, to determine that the cost of the standard had a reasonable relationship to the benefits. This time the Supreme Court rejected the argument that the Occupational Safety and Health Act required the use of cost-benefit analysis. The Court said, "Congress itself defined the basic relationship between costs and benefits, by placing the benefit of workers' health above all considerations save those making attainment of this 'benefit' achievable. . . . Thus cost benefit analysis by OSHA is not required by the statute because feasibility analysis is."[39] The Supreme Court decided that neither the language of the act nor its legislative history indicated that Congress meant for the agency to engage in cost-benefit analysis. In the benzene decision, the court had considered feasibility secondary to the requirement of proving significant risk. In the cotton dust decision, the court said that besides feasibility analysis Congress did not contemplate any further balancing by the agency for toxic or harmful agent standards. The mixing of the scientific activity of assessing risk and the sociotechnical activity of judging risk caused much of the bitterness among groups associated with setting the vinyl chloride standard and made risk assessment vulnerable to political pressure.

Decision making for regulating chemicals has been evolving since the 1970s. A critical concept now utilized is that of risk. The relatively new activities of risk assessment and risk management were at the heart of the vinyl chloride controversy.[40] Increased pressure was placed on regulators to justify proposed standards. Because of insufficient scientific data to assess risk and the absence of inference guidelines to utilize incomplete data to derive a risk assessment, the final recommendations inextricably mixed risk assessment and risk management efforts to describe the risk. Assessment and management of risk got mixed up with the debate on how to arrive at an acceptable risk, all of which pointed to the need for a more rational method to estimate risk and the desirability of clearly differentiating risk assessment, a scientific activity, from risk management, a political activity subject to additional factors, including economics. Today we are moving toward the separation of assessment and management of risk.

Controversy still exists over how to assess risk. Many of the scientific questions of 1974 are unresolved. We still have not determined what constitutes a precise level of exposure for a carcinogen, whether a safe level of exposure exists, or whether animal dose-response data can be extrapolated to estimate the adverse effect in human populations. In the realm of policy there is still philosophical controversy over what is just and fair, over balancing the cost of controls against the risk, and how certain one must be about the data before considering the extent of regulation. These questions still remain unanswered. Scientific data do not decide the bottom line. Even after the database is sufficiently complete to provide a scientifically defensible assessment of risk in any particular case, the risk management dialogue will continue to involve competing social values. At this point, the wisdom of clearly differentiating risk assessment and risk management is clear, but whether this is entirely possible or, if it is, whether it will mute the intense controversies in this area is not clear. The only consensus we have on controlling carcinogens in the workplace is on the methods of control. In other words, while there still exists controversy over assessment and judgment of risk, we can agree on workable, effective management methods to reduce the impact of carcinogens in the work-place. Our state of knowledge determines the control measures we apply. In the case of workplace-induced cancer, the major means today, as it was in 1974, is to minimize exposure to chemicals capable of producing cancer.

References
1. Doniger, D. *The Law and Policy of Toxic Substances Control: Resources for the Future*. Baltimore: Johns Hopkins Press, 1978.
2. Lowrance, W. *Of Acceptable Risk*. Los Altos, CA: William Kaufmann, 1976.
3. National Research Council. *Risk Assessment in the Federal Government: Managing the Process*. Washington: National Academy Press, 1983.
4. Environmental Protection Agency. *Scientific and Technical Assessment Report on Vinyl Chloride and Polyvinyl Chloride*. 600/6-75-004 Washington: EPA Office of Research and Development, 1975.
5. Ibid.
6. Ott, G., R. Langner, and B. B. Holder. "Vinyl Chloride Exposure in a Controlled Industrial Environment." *Archives of Environmental Health* 30(1975): 333–339.
7. Patty, Frank A. ed. *Industrial Hygiene and Toxicology*. 2d ed., Vol. 2. New York: John Wiley & Sons, 1962.
8. Milby, Thomas H., ed. *Vinyl Chloride: An Information Resource*. NIH 79-1599 Washington: U.S. Department of Health, Education, and Welfare, 1978.
9. Marstellar, J. H. and S. Juhe. "Chronic Toxic Liver Damage in PVC Production Workers." *Geriatric Medical Journal* 98(1973): 2311–2314.
10. Selikoff, I. and H. E. Cuyler, eds. "Toxicity of Vinyl Chloride–Polyvinyl Chloride." *Annals of the New York Academy of Sciences* 246(1975): 6.
11. Viola, P. L., A. Bigotti, and A. Caputo, "Oncogenic Response of Rat Skin Lungs and Bones to Vinyl Chloride." *Cancer Research* 31(1971): 516–519.
12. Report of Committee on Government Operations, House of Representatives. *Chemical Dangers in the Workplace*. #94-1688 Washington: U.S. Government Printing Office, 1976.
13. Ibid., 4–5.
14. American Conference of Government Industrial Hygienists. *Threshold Limit Values for Chemical Substances in the Work Environment*. 1983–84.
15. Committee on Government Operations, *Chemical Dangers*, 16.
16. Ibid., 17.
17. Hearings Before a Subcommittee of the Committee on Government Operations, House of Representatives, *Control of Toxic Substances in the Workplace*. Washington: U.S. Government Printing Office, 1976.
18. Ibid., 181.
19. Creech, J. L. and M. N. Johnson. "Angiosarcoma of Liver in the Manufacture of Polyvinyl Chloride." *Journal of Occupational Medicine* 16(1974): 150–151.
20. Makk, L., J. Creech, J. Whelan, and M. Johnson. "Liver Damage and Angiosarcoma in Vinyl Chloride Workers: A Systematic Detection Program." *Journal of the American Medical Association* 230(1974): 64–68.
21. Heath, C., H. Falk, and J. Creech. "Characteristics of Cases of Angiosarcoma of the Liver Among Vinyl Chloride Workers in the United States." *Annals of the New York Academy of Sciences* 246(1975): 231–236.
22. Letter from Peter Bommarito, Chairman, Industrial Union Department (AFL-CIO) Committee on the Health and Safety of Workers. Exhibit 3, Vinyl Chloride Hearing, July 1974.

23. The United States Department of Labor. Occupational Safety and Health Administration. "New Standard for Vinyl Chloride." *Job Safety and Health* (July, 1974).
24. OSHA Hearings on Proposed Standard for Vinyl Chloride. Exhibit 7 (hh) 1974. Preprint of ICS Series 322, Maltoni, Occupational Carcinogenesis.
25. Selikoff and Cuyler, "Toxicity," 5.
26. 39 FR 12341.
27. OSHA Hearings on Proposed Standard for Vinyl Chloride. Exhibit 20D, 1974, statement of Anton Vittone, president of B.F. Goodrich Company.
28. FR 25892.
29. *Chemical and Engineering News* 52(July 1974): 26.
30. OSHA Hearings on Proposed Standard for Vinyl Chloride, Exhibit 22(F), 1974, statement of Vince P. Ficcaglia, manager of economic analysis and forecasting, Arthur D. Little.
31. *Business Week* 2336(June 1974): 30.
32. 39 FR 35892.
33. 45(15) FR 5010.
34. Office of Technology Assessment. *Preventing Injury and Illness in the Workplace.* OTA-H-256 Washington: U.S. Congress, Office of Technology Assessment, 1986.
35. Ibid., 231.
36. Corn, J. K. and M. Corn, "The History and Accomplishments of the Occupational Safety and Health Administration in Reducing Cancer Risks." In *Reducing the Carcinogenic Risks in Industry,* edited by Deisler, P. F. New York: Marcel Dekker, 1984.
37. Rothstein, Mark A. *West's Handbook Series, Occupational Safety and Health Law,* 2d ed. St. Paul: West Publishing Co., 1983.
38. Ibid.
39. Ibid., 85.
40. National Research Council, *Risk Assessment.*

Chapter 8

Cotton Dust

Byssinosis, an occupational respiratory disease caused by exposure to cotton dust, has been until recently either ignored, misunderstood, or its existence actively denied. Association between dust and respiratory illness appeared early in the history of the cotton textile industry, but the definition and identification of the respiratory disease so closely associated with cotton dust and the needed action to prevent dust-induced disease lagged far behind the growth of the textile industry.

For over a century byssinosis was known, but its recognition as an occupational disease did not occur in the United States until the 1960s. Before then, in the United States, the relationship between cotton dust and byssinosis remained largely unexplored and when considered, if indeed considered at all, was regarded as an uncommon disease.

Epidemiological studies of the 1960s showed the high incidence of byssinosis in the textile industries in the United States and highlighted the need to take action to prevent this disease from occurring as the result of work in textile mills. After the epidemiological studies of the 1960s, a debate ensued over policy toward the disease. Controversy embroiled the textile industry, labor, government agencies, researchers, and lawmakers. Because of political and ideological polarization, the movement for adequate and reasonable legislation to protect cotton workers at risk did not go smoothly.

The controversy brought into focus the altering perception of hazard and the continually changing definition of risk. Perplexing decisions affected textile workers' lives and involved millions of dollars in expenditures to alleviate hazardous conditions in the cotton mills and to compensate byssinosis victims. Policy makers confronted with these decisions raised the question of the price we are willing to pay for protecting health. For example, do the benefits of health protection warrant the expenses of an effective standard?

The purpose of this chapter is to present a historical perspective on the disease byssinosis in order to illustrate the impact of scientific, sociopolitical, and economic inputs to health policy decisions. The contemporary history of byssinosis illustrates that risks we once judged acceptable are being reevaluated in the light of new scientific evidence, technology, and changed social attitudes.

DESCRIPTION OF BYSSINOSIS

Byssinosis, commonly called *brown lung,* occurs in the presence of cotton, flax, or hemp dust. Its initial acute symptoms are tightness in the chest, dyspnea, and a cough after returning to work on Monday. A decrease of ventilatory capacity occurs during the workday. Later, symptoms extend to other workdays. Eventually, there is a chronic stage, resulting from continued exposure over a long period of time, with severe continuous dyspnea, chronic cough, and permanent ventilatory insufficiency. Irreversible and disabling chronic obstructive pulmonary disease is referred to as COPD. Total disability and death often follow. The acute effects of exposure may be reversed, but long-term chronic effects cannot be reversed.

Richard Schilling devised a grading system for byssinosis, which is used as the standard clinical method for classifying byssinosis based upon symptoms. The degree of byssinosis is classified as follows:

Grade 0: no evidence of Monday chest tightness or breathing difficulty
Grade 1/2: occasional chest tightness on the first day of the week
Grade 2: chest tightness every first and other days of the working weeks
Grade 3: grade 2 symptoms accompanied by evidence of permanent incapacity from diminished effort intolerance and/or reduced capacity.[1]

Functional detection of byssinosis can be made by having a worker blow hard into a spirometer (a device used to measure lung volume changes) on Monday morning and before he or she goes home at the end of a shift. The spirometer measures forced expiratory volume in one second ($FEV_{1.0}$). Most persons with chest tightness show a decrement in expiratory flow rates after six hours of dust exposure, after two or more days without exposure. Flow measured at 50 percent of vital capacity is sensitive to small changes in airway capacities, notwithstanding that $FEV_{1.0}$ has been measured in most studies of byssinosis. It has been suggested that measurement of $FEV_{1.0}$ and FVC (forced vital capacity) before and after six hours of exposure to determine whether exposure caused a decrease should be included in surveys for byssinosis. Most individuals with grade 1, 2, and 3 byssinosis have a moderate to marked decrease in $FEV_{1.0}$ after six hours of dust exposure. However, evidence of no decrement in $FEV_{1.0}$ does not preclude the diagnosis of byssinosis in persons with symptoms.[2]

Perhaps a cotton worker's description of this illness can give insight into the nature of byssinosis. In 1970, Mr. Lacy Wright, a textile worker with byssinosis, described his illness before a hearing of the Senate Subcommittee on Labor on the Occupational Safety and Health Act. This subjective personal account of illness should be added to the objective, clinical description of the disease in order to describe more fully the effects of byssinosis.

I have worked for all of my life in the cotton mill, for 44 years. . . . When I went to work I was in good health, I thought, as a boy of that age. I could get out and run and play and wasn't bothered any. So I went to work in the carding department.

The mill then was much more open than today. But I began to notice when I would play, I couldn't breathe as good as I could before I went to work. . . .

So I kept on working and my breathing kept on getting worse, I went along, and I got to the point that I would just cough and sometimes hang over a can and cough and become nauseated in my stomach, and I began to be bothered about my condition. . . .

I got to the point where I just couldn't breathe in there at all, and I went to the doctor, approximately three or more years ago. . . .

He said, "It is a lung condition in your lung."

I said, "Is that why I'm so short of breath?"

He said, "It sure is."

I said, "Well if I don't get some relief, I am not going to be able to work."

He said, "Is it dusty where you work?"

I said, "It sure is."

He said, "There are just two things you can do then."

I said, "What is that?"

He said, "You can get out of that dust, or you can stop breathing."

I looked at him and I realized I had gotten worse and worse, year after year. I remembered when I could breathe fairly well, and I had gotten to where, when I got home, I couldn't get out of my car and into the house after a while. The doctor said I just had to quit.

All of my life, I have been very active, but now I am short of breath, and I can't do anything. I get so short of breath and so weak I can hardly go.

As far as I see things, and as far as the doctor says, I will be like this as long as I live.[3]

AT RISK

Because byssinosis occurs in the presence of cotton dust, at risk in particular are workers in cotton mills, but others involved in the processing of cotton from ginning to finished goods can manifest symptoms. In cotton mills the areas of preparation—opening, cleaning, picking, and carding—have a higher incidence of byssinosis. A higher incidence of byssinosis occurs in workers with more than 18 years' exposure. Males are affected twice as often as females, and smokers more than nonsmokers.[4]

In 1979 epidemiological evidence and exposure level data enabled the U.S. Department of Labor to estimate that about 84,000 active workers currently suffer from byssinosis of all grades. Table 8-1 indicates the number of workers exposed to cotton dust and the expected number of byssinosis cases by industry.[5]

Bouhuys and colleagues used a community study of cotton textile workers to estimate that 30,000 active and retired textile workers suffer totally disabling byssinosis. He estimated that because textile workers are employed for an average of 35 years, about 800 of these workers would become totally disabled each year.[6]

COTTON MANUFACTURE IN THE
UNITED STATES

Cotton cultivation and its manufacture into textiles and other products are important agricultural and manufacturing industries in the United States. Major cotton-producing states are California, Texas, Mississippi, and Arkansas, although at least 19 states grow cotton commercially. By the second half of the twentieth century, most cotton textile mills had relocated from New England to a small area on the southeastern seaboard. In 1966 New England cotton mills comprised approximately 3 percent of the total textile production in the United States, and the Southeast accounted for approximately 94 percent of the total. Between August 1, 1972, and July 31, 1973, 13.2 million bales (about 3.3 million tons) of cotton were produced in the United States. About 74 million bales were processed in the United States, and the number of workers involved in cotton fiber processing in 1973 was estimated at 800,000. Textiles is one of the largest manufacturing industries in the United States. It employs about 1 million workers. The cotton textile division employs 250,000 workers, more than 90 percent in the southern states.[7]

In the United States cotton is harvested mechanically (spindle picker or striper). The ginning process mechanically separates fibers from seeds. The fiber or lint is then compressed into bales and transported to cotton

TABLE 8-1. Workers Exposed to Cotton Dust and Expected Byssinosis Cases, 1977

Industry	SIC[a] Number	Exposed Workers (Thousands)	Assumed Byssinosis (All Grades) Prevalence (%)	Expected Byssinosis (All Grades) Cases (Thousands)
Agricultural total		107.7		18.26
Cotton ginning	0724	92.6	17	15.74
Classing	9641	0.3	15	0.05
Cotton compresses and warehouses	4221	10.8	15	1.63
Cottonseed oil mills	2074	4.0	21	0.84
Yarn manufacturing total		188.1		43.74
Broad woven fabric mills, cotton	2211	86.6	25.8	22.34
Broad woven fabric mills, fiber and silk	2221	47.0	15.5	7.29
Circular knit fabric mills	2257	1.0	25.8	0.30
Yarn spinning mills	2281	34.4	25.8	8.88
Texturizing, throwing, twisting, and winding mills	2282	13.9	25.8	3.59
Thread mills	2284	3.6	25.8	0.93
Tire cord and fabric	2296	1.6	25.8	0.41
Fabric manufacturing total		230.2		17.9
Broad woven fabric mills, cotton	2211	16.2	16	2.59
Broad woven fabric mills, artificial fiber and silk	2221	8.8	7.1	63
Narrow fabrics	2241	4.5	14	0.63
Knitting mills	225	200.7	7	14.05
Textile waste		33.7		3.71
Mattresses and bedsprings	2515	25.1	11	2.76
Total		559.7		83.61

[a] SIC = Standard Industrial Classification

SOURCE: U.S. Department of Labor Report to the Congress, *Cotton Dust: Review of Alternative Technical Standards and Control Technologies*, May 14, 1979.

mills. At the mills the cotton is prepared (opening, cleaning, and picking) for carding, and the cotton is cleaned. During carding, fibers are separated and formed into a bundle for further processing. The carded fiber, in the form of loosely assembled rope, is elongated, straightened, and decreased in diameter, as it is stretched in the drawing operation. Fibers are combed. The spinning process twists fibers together to make yarn that is wound on spools. The yarn is then woven into cloth.

EARLY ASSOCIATION BETWEEN EXPOSURES TO COTTON DUST AND RESPIRATORY ILLNESS

Cotton cultivation and manufacture into textiles occurred in ancient India, but the mechanization of cotton textile production did not take place until the eighteenth century in England, when the new inventions, new sources of power, and innovations in the organization of production that characterize the Industrial Revolution first occurred in the manufacture of cotton textiles. In the United States, early mechanization of industry is also associated with cotton textiles. Cotton mills in both countries dominated political, economic, and social life.

Association between exposure to cotton dust and the development of acute or chronic respiratory illness was made even at the beginning of mechanization of the cotton industry. With machines came dust, and descriptions of abnormal conditions of the respiratory tracts among cotton workers appeared as early as 1832.

Dr. Kay, a resident of Manchester, England, who worked at the Ardwick Dispensary, described "spinners' phthisis."

In many cases which have presented themselves at the Ardwick and Ancoats Dispensary, the disease induced has appeared to me to differ from ordinary chronic bronchitis. In the commencement of the complaint, the patient suffers a distressing pulmonary irritation from the dust and filaments which he inhales. Entrance into the atmosphere of a mill immediately occasions a short, dry cough, which harasses him considerably in the day, but ceases immediately after he leaves the mill and inspires an atmosphere free from foreign molecules. These symptoms became generally more severe; the cough is at length very frequent during the day, and continuous even after its employments have ceased, disturbing the sleep, and exhausting the strength of the patient, but it is accompanied with little or no expectoration. In this stage, he seeks medical aid. He is harassed with a frequent cough, which is often excited by speaking, slight exertion, or change of temperature. The patient sometimes expectorates a little, but the cough is often dry and short, and recurs incessantly. He experiences a diffused and obscure sensation of uneasiness beneath the sternum; in sudden exertion a pectoral oppression ensues, arising as it were from an inability to dilate the chest freely in the ordinary inspirations. The whole respiratory system evinces a great and early excited irritability. There is little febrile action. On the application of the stethoscope, no rales are in general perceptible, the respiratory murmur is scarcely puerile. The patient is easily afflicted with acute bronchitis on exposure to its exciting causes and this disease often succeeds the previous complaint.[8]

American references to mill workers' health in the nineteenth century also exist. Dr. Josiah Curtis attempted to discover the effects of the factory on the health of workers. In 1849 he noted the effects of poor ventilation and high temperatures on the health of cotton workers.[9] At a hearing for a ten-hour day before a legislature committee in 1845 in Massachusetts, witness Elizabeth R. Hemingway said,

The working time was too long; the meal times too short. The air in the factory was bad; over 150 persons worked in one room, for example, where 293 small lamps and 61 large ones burned morning and evening during the winter months. There was no day when fewer than six girls stayed away from work because of illness; as many as 30 had been known to remain at home on one day for that reason. In all seasons flying lint was a source of great discomfort.[10]

Health problems related to working conditions in cotton mills were also noted in Cohoes, New York. A report in the *Medical Press* in 1890 states, "Bronchitis was most pronounced among cotton mill hands."[11] The article's author suggested that operatives use a respirator made of cotton wool over their mouths.

The Thirteenth Annual Report of the Massachusetts Bureau of Statistics of Labor included a discussion of conditions relating to health in cotton mills. "Cotton fibers [in] constant circulation through the air get into the throat and give everyone the appearance of having a cold."[12]

It should be noted that in general, working conditions in the New England mills of the nineteenth century were bad. Long and arduous hours of work, dangerous working conditions, low wages, and overcrowded housing conditions were more of an issue than health. Health took a long time to emerge as an issue. Perhaps because other problems were more immediate and easier to define, mill workers would wait until the twentieth century to seek healthful working conditions.

Nevertheless, even in the nineteenth century, there was an association between sickness and working in the cotton mills, however ill defined.

Those associated with cotton textile mills recognized that abnormal conditions of the respiratory tract occurred among cotton operatives.

In 1863 an article published in *Lancet*, a British medical journal, described the health effects of cotton dust upon operatives in a mill.

> The first process the raw cotton undergoes is the mixing of one staple with another. Much dirt and dust is disengaged in this operation. The respiration is affected from the dust irritating the respiratory passages of the mixer, and coughing and sneezing are the frequent consequences, which disengage from the bronchial membrane a quality of salty colored expectoration, which when placed under the microscope, is seen to consist of very fine short fibres of cotton in air bubbles and mucus. . . . After passing from the mixers, the cotton passes through the hands of the willowers or scutchers. When ventilation is not assisted by ventilating chimneys of tin or wood, which takes off more effectually the dense atmosphere with which the rooms are charged, the willowers and scutchers suffer in the same manner as the cotton mixers. From the immense velocity of the machines used, the revolutions being 1500 per minute, the quantity of short fibres of cotton set afloat in these rooms is very great. It would be difficult to recognize a man at twelve yards distance from the density of floating fibres, modified of course, very much by a wet or dry day. The strippers, grinders, and cardroom hands are engaged in the next process of cotton manufacture. They mostly suffer from a spasmodic cough, sore throat, expectoration of blood, pneumonia, and confirmed asthma, with oppression of the chest. . . . A carder seldom lives in a cardroom beyond forty years of age. Many have to give up working much younger. Forty-five to fifty years are the average ages.[13]

EARLY INVESTIGATIONS INTO THE NATURE OF ILLNESS ASSOCIATED WITH COTTON MANUFACTURING

Once it was recognized that abnormal conditions of the respiratory tract occurred among cotton operatives, it was then necessary to show that abnormal conditions of the respiratory tract occurred more often among cotton operatives than among the general population and that the abnormal conditions and illnesses could be attributed to the inhalation of cotton dust. It remained then to describe the etiology of the disease caused by inhalation of cotton dust. Thus, in order to begin to control the respiratory illness caused by cotton dust, it had to be shown to exist, described, named, and its etiology understood. There is evidence that investigations into the nature of disease caused by cotton dust took place and that a long-standing and worldwide interest in the effects of exposure to cotton dust upon the respiratory tract existed early in the twentieth century. In

1947 the United States Public Health Service published a review of litera-ture related to affectations of the respiratory tract in cotton workers.[14] It contained 267 references to cotton dust and health in a variety of languages and covered the period 1827 to 1947. References also contained informa-tion on amounts of dust in the air, sizes of dust particles, and chemical analysis of dust.

Table 8-2 consists of 36 surveys of cotton workers' health from 1827 to 1946. It includes the investigator, country, year, and major findings of each of the studies. The early literature on cotton workers' health stresses occupational hazards as well as other cotton workers' problems that could affect health, including low wages, inadequate diet, poor housing, and sanitation.

Thus, references to respiratory disease associated with cotton manufac-ture existed from the beginning of the history of mechanization of that industry. It took time to name the disease caused by cotton dust. Proust first used the term *byssinosis* in 1877. Before that and well into the twentieth century, observers gave the disease many different names.

Two English studies are worthy of note. The first, in 1908 by Collis, examined 126 men employed in Blackburn, England, and found 73.8 per-cent complaining of or suffering from an asthmatic condition due to inhala-tion of dust. Collis noted the association between length of employment and prevalence and described the symptoms of byssinosis.

The course of the trouble caused is as follows: As soon as the individual begins to suffer, he finds his breathing affected. On Monday morning, or after any interval away from the dust, on resuming work he has difficulty getting his breath. This difficulty is worse the day he comes back. Once Monday is over he is alright for the week. . . . The man usually gets "tight" or "fast" in the chest, and he finds difficulty in filling his lungs; to use his own expression, "the chest gets puffed up." Consequently, he becomes thin in the face and body. As the chest trouble develops into a typical form of asthma, the action of the diaphragm becomes less and less effective, until the only action of this great respiratory muscle is to fix the lower ribs; at the same time the superior intercostal muscles are being brought more and more into use, and the extra ordinary muscles of respiration are more and more called into play to carry on the ordinary act of breathing. The sternum becomes more prominent, and the chest becomes barrel shaped. Meanwhile, the extra tax thrown on the lungs leads to some degree of emphysema. There is little or no sputum produced, and what little there is is expectorated with difficulty. It is not infrequently stained with blood, but I only found doubtful physical signs of phthisis in one case who so complained.[15]

TABLE 8-2. Surveys of Cotton Workers' Health, 1827–1946

Investigator	Year	Country	Finding
Gerspach	1827	France	Beaters and pickers work 15–16 hours per day in thick dust; cough and expectoration of blood; many cases designated as tracheal phthisis
Thackrah	1832	England	1,500 Employees in Manchester mill; found children in carding and preparing room puny, frequent catarrh and moderate cough, mild disorders of nervous and digestive systems; found occasional bronchitis and pulmonary maladies
Vilerme	1840	France	Studied economic and social factors affecting workers; cotton beaters first affected by dryness of mouth, then progressive cough, slow and serious chest disease; illness could be alleviated by removal from workroom; disease resembled pulmonary phthisis
Mareska & Heyman	1845	Belgium	Beater picker and beater-opener room most unhealthy; pulmonary affectations among cardroom workers most prevalent; phthisis, laryngitis also prevalent; workers complained dust inconvenienced them most on Mondays and Tuesdays
Bredow	1851	Russia	Large number of workers suffered from pulmonary tuberculosis
Schuler	1872	Switzerland	Dust present particularly in air of cardroom, also in spinning and weaving rooms; skin diseases, especially eczema and varicose veins; chronic pneumonia, emphysema, asthma, and blennorrhea prevalent
Erisman	1887 1888	Russia	Prolonged work in factory harmful to growth and development
Verhoege	1904	France	Of 1,845 cotton, wool and tow workers, 28% reported chronic bronchitis; those with longer occupational histories more often affected.
Pierotti	1906	Italy	Dust mentioned as cause of illness (180 men and 620 women); 10% of workers had chlorosis; bronchitis, pleuritis, and pneumonitis in 21.2% of men and 8.6% of women.
Collis	1908	England	Strippers and grinders examined in cardrooms of 31 mills in Blackburn (126 men); after years of work, strippers and grinders develop progressive respiratory distress, complaints of chest tightness; symptoms more severe on Monday; emphysema of varying degree; 73.8% affected.
Preis	1908	England	Sick benefit records analyzed for approximately 5,700 male and female spinners and weavers (1905–1909); tuberculosis, alcoholism, diseases of cochlea, or rare infection of inner ear noted

Safford	1916	U.S.	Of 691 boys under 18, average weight, height, and chest size of boys reported below average for other groups of boys; some dulling of hearing, functional disorders of the heart; no conditions of the eyes, nose, mouth, teeth, tongue, lungs, skin, or digestive tract that could not be attributed to causes other than occupational
Schmidt	1924	Germany	449 People 25–55 years old, cotton spinners, farmers and their dependents, students; spinning factory workers inhaled dust; cotton dust considered to be pathogenic factor causing catarrh of upper respiratory tract; roentgenologically, persons active in spinning factories for decades showed visible large and moderate-sized bronchi as signs of mucosal proliferation and thickening of the walls in conditions of chronic irritation; presence of pneumoconiosis not established
Schilling	1925	Germany	Three hundred spinners and 150 agricultural workers. X-rays, some autopsies. Constant exposure to cotton dust. All spinners had history of frequent pharyngeal catarrh and bronchitis. In workers with over a decade of experience bronchitis occurred with more or less emphysema. Concluded inhalation of cotton dust resulted only in a heightened disposition to acute and chronic bronchitis and consequently an early emphysema of the lung. No heightened disposition to tuberculosis (used X-ray).
Bekritzky	1927	Russia	400 Workers in carding and combing rooms with dust concentrations of 14.2 mg per cubic meter of air; dust in slubbing and scrutching rooms; in rooms with much dust 60% of workers had atrophic rhinopharyngitis, 24% had atrophic laryngitis; no mention of pulmonary disease
Report of Chief inspector of factories	1927	England	1,011 Weavers inspected for deafness
Socolow	1928	Russia	Judged health of 758 workers, one visit; morbose alterations considered as probably due to the profession; found diseases of respiratory tract, particularly nose and pharynges
Bianchi	1932	Italy	Of 2,600 workers (900 in cotton manufacture, the rest jute and hemp), 320 had respiratory affections
Caso	1932	Italy	600 Workers, age 15–65 years, in a spinning mill; 60% of women carding, spinning, and weaving department in dusty atmosphere; cardroom workers had conjunctivitis and blepharitis; 100 workers had disturbances of respiratory tract; 50 workers age 15–56 complained of dryness of throat, difficulty in swallowing, sneezing, and mucoid rhinitis; improvement occurred on removal from work; 20 persons who worked 5–10 years showed evidence of chronic pulmonary pathology not easily classified clinically

TABLE 8-2. Surveys of Cotton Workers' Health, 1827–1946 (*Continued*)

Investigator	Year	Country	Finding
Bokser and Ryabova	1932	Russia	324 Workers with control group. 82 men, none having worked in a dusty occupation; study included determination of dust exposures; group with 10–20 years experience complained of shortness of breath, either at rest or with slight exertion; persons 45–50 years old had evidence of emphysema and reduction of chest expansion
Koelsch	1932	Germany	13 Men and 21 women cotton workers: 62% 20–40 years old, 75% worked more than 10 years, most for 10–25 years; indications of cardiac hypertrophy in 13 of 34; also found bronchitis and emphysema
Report of Department committee	1932	England	Government inquiry into diseases of cardroom workers; evidence of problem of respiratory disease among card strippers and grinders
Cano	1932	Spain	Cotton and woolen factories; much dust produced in all operations; affectations of respiratory tract, exclusive of tuberculosis, greater than average; mortality from tuberculosis increased and mortality from heart disease took second place
Koelsch	1933	Germany	Of 27 men and 48 women, 87% had over 10 years in the cotton industry, some as long as 45 years; 17 men and 25 women complained of chest difficulties; all beaters and carders complained of temporary asthmatic pain in throat and chest on Mondays, pains disappear by Thursday; 34 of 75 had heart enlargement
Britten, et al.; Bloomfield and Dreessen	1933	U.S.	Males in one cotton plant: 749 weavers, 438 carding, 441 spinning; dust determinations made; carding, spooler, and picker
Klionskii and Yapolskii	1934	Russia	Of 18 males and 98 females working in dusty atmospheres, 82% complained of cough, 70% of shortness of breath, 17% history of hemoptysis, 57% bronchitis, 38% of cardiac hypertrophy; it was concluded that textile dust does not provoke the classical pneumoconiosis
Beintker	1934	Germany	Reports 29 cases of respiratory disease in textile mills of Westphalia
Prausnitz	1934	England	Attempt to determine cause of respiratory disease among cardroom workers
Chilla	1934	Italy	Investigation of health and sanitation of approximately 3,000 workers; dust in opening, mixing, and beating rooms; laryngitis, pharyngitis, and tonsillitis among workers, a few with bronchopulmonary affections
Thiry	1938 1939	Belgium	125 Workers in a cotton spinning mill; dust especially heavy in mixing and beating rooms, less dust at the cards; 24 workers at summary, 6 at mill for a few months,

158

others 10 years; those working a few months and 1 working 8 years had normal x-rays; 10 or more years, 2 had normal lungs, 8 had pulmonary sclerosis with emphysema, 2 had chronic bronchitis, and 2 had pachypleuritis

Trier	1938	Denmark	262 Cotton workers, 81 in spinning, 181 bale breaking
Thiry	1940	Belgium	301 Workers, 80 in preparatory processes, 221 in spinning; farmer exposed to more dust; cervico-axillary adenitis in 74% of workers in preparation; 58% of spinners hypertensive; respiratory capacity of many workers classified as weak or very weak; 6 cases of bronchitis, 2 of asthma, 8 of tonsillitis, 1 endocarditis, and 2 of exophthalmic goiter
Smith	1942	U.S.	156 Men and women in cotton batting; chest x-rays, 51 men and 21 women physical exams; cotton dust flew freely about and reached high concentration; chest x-rays revealed 1 cardiovascular disease, 2 lung conditions unrelated to dust, and 1 possible case of tuberculosis; physical exams revealed varying degrees of upper respiratory infection, 2 with basal rales in lungs and 2 had diminished breath sounds
Forero and Thomas	1943	Chile	282 Textile factories (cotton, silk, rayon, etc.), approximately 8,770 workers; hygienic conditions and protective devices deficient in most factories; byssinosis, chronic laryngeal tracheitis, chronic bronchitis stated to be present
Ritter and Nussbaum	1945	U.S.	12 Men age 37–70 with asthma precipitated by exposure to cotton dust given physical exams and x-rays; 26 persons who worked more than 30 years in carding or picking in textile mills or lint rooms and seed storage bins of cotton mills: "No evidence was found in a cross section of study of the Mississippi cotton industry to support the existence of byssinosis as a clinical entity arising among employees exposed to upland cotton"
Roman	1946	Canada	Based conclusions on life histories of 205 textile industry employees (executives, office men, cotton sorters, spinners, weavers, dyers, bleachers, nappers, mechanics, painters, and carpenters); Conclusions: "The author considers that there is no industry with less hazard and less occupational disease than the cotton textile industry; tuberculosis, bronchitis, and asthma formerly associated with the industry have been eliminated by means of improved lighting, scientific humidification and better ventilation with regard to clean dust free air"

SOURCE: G. H. Camanita, et al. *A Review of the Literature Relating to Affections of the Respiratory Tract in Individuals Exposed to Cotton Dust.* Public Health Bulletin 297 (Washington: Federal Security Agency, U.S. Public Health Service, 1947), 12–23.

Hill performed the second study. He inquired into sickness among cardroom operatives compared with other cotton operatives in the same spinning mills and made the following conclusions:

1. Strippers and grinders, the cardroom workers exposed to the maximum amount of dust, and other cardroom workers have in comparison to other operatives in the cotton industry and in comparison with all males in England and Wales high death rates from respiratory diseases.
2. From all causes of sickness, experience is divided into respiratory and nonrespiratory causes. Strippers and grinders and other cardroom workers have somewhat less sickness than other male workers from all causes of sickness excluding respiratory diseases. After 30 there is a pronounced excess of illness from respiratory diseases. Rates are two to three times as high as rates found for other workers in the mills.
3. A consideration of male and female workers who at one time worked in the cardroom, but were not in it during the period of inquiry, showed those workers had unduly high rates from respiratory illness.
4. The excess of respiratory illness among cardroom workers cannot be easily explained by any other factors except environment.
5. There is no evidence of an excess of respiratory illness among the younger workers who have worked only under modern conditions. This is not conclusive that modern conditions do not cause respiratory illness because it may be that some years of exposure are necessary before the effect of environment will be shown.

It seems evident that the conditions in the cardroom before localized exhaust ventilation and vacuum for striping were introduced were distinctly unfavorable to health, both men and women being affected. Such extreme conditions no longer exist and it is to be hoped that the injurious effect upon health has changed with the change in environment. In the absence of positive evidence that this is so, the reduction of the operatives' exposure to dust and fibre, to the maximum extent possible, is obviously desirable.[16]

In 1936 Prausnitz introduced the term *Monday fever* into the literature on byssinosis.

After working for years without any appreciable trouble except a little cough, they noticed either a sudden aggravation of their cough, which becomes dry and exceedingly irritating, or peculiar attacks of breathlessness. These attacks usually occur on Mondays, while the rest of the week finds them in fairly good condition. For a long time, the trouble may be almost or entirely limited to this *Monday fever,* but gradually the symptoms begin to spread over the ensuing days of the week; in time the difference disappears, and they suffer continuously. When ultimately they are forced to give up work, some improvement may occur. But if they resume work, they are forced sooner or later to give it up again. In some this is most characteristic: A rest of some weeks or months gives them an opportunity of recovering so far they feel perfectly fit, yet on the day that they return to the mill, they are struck down with the disease and are hardly able to return home unaided.[17]

Textbooks such as *Diseases of Occupation* included byssinosis as an occupationally caused disease: "There are various forms of dust diseases of the lungs or pneumokonioses, e.g., anthracosis, due to coal dust; chalicosis and silicosis, due to sandstone and mineral grit; sialerosis, due to iron; and byssinosis, to cotton fibers."[18]

Studies of cotton mill workers also ascribed incidence of illness to socioeconomic conditions and to other harsh working conditions such as low wages, long hours, endurance of heat, and humidity.

By 1940 statistical information on the health of cotton workers in England indicated a high rate of mortality from respiratory disease. Reports emphasized that high mortality from respiratory disease occurred chiefly among cardroom and blow-room hands, strippers, and grinders. Mortality from respiratory diseases was considered due to dust rather than heat and humidity.[19] Although the mode of action of the dust was not completely understood, evidence existed to indicate that dust could act as a mechanical irritant, as a course of microbiological toxins, of histamine, and of allergens.[20] Thus, by 1940, a literature existed pertaining to respiratory tract effects among workers exposed to cotton dust. From that literature, the following generalizations can be made as to what was known about byssinosis: It had been named and described. Although cotton dust was not the only hazard in the textile mills, it was assumed to be a principal one with regard to respiratory tract ailments. The early stages of processing (ginning, bale-breaking, picking, and carding) involved greater dust hazard than spinning and weaving. Byssinosis occurred in workers with long and continuous exposure to cotton dust. Byssinosis was observed in all countries with a cotton textile industry, and the literature on the subject of health of cotton operatives was international, with the British work most complete. Byssinosis represented the end result of a chronic process.

Complete data did not exist relative to the quantities of dust in the atmosphere of cotton factories or the physical, chemical, and biological composition of cotton dust. In 1940, the British passed legislation recognizing byssinosis as a compensable industrial disease.

THE AMERICAN EXPERIENCE:
TARDY RECOGNITION

Until the 1960s, in spite of its widespread recognition, the problem of serious dust disease among cotton workers was hardly known to exist in the United States, a nation with a large textile industry.

In order to comprehend American tardiness in recognizing the existence of byssinosis as an industrial disease, it is necessary to understand subsequent industrial relationships in mill towns of the southeastern United States and a series of developments over decades.

Textile manufacturing is one of the largest industries in the United States and employs almost 1 million workers. Its largest division, cotton textiles, employs approximately 250,000 workers. More than 90 percent of the manufacturing occurs in southern states. Cotton mills dominated the political and economic life of the New England towns, where the industry originated, and continued to do the same in the South, where they eventually relocated at the end of the nineteenth century, ushering the Industrial Revolution into that region.

The factors that favored the growth of new industry did not always help working people. Indeed, the availability of a supply of cheap and tractable labor in the South is such an example.

> Competition for jobs was so severe that wages were minimal, with the manufacturer furnishing housing and many other facilities for the maintenance of life. Additional thousands of tenant farmers and mountaineers stood ready to take the jobs of mill workers who, through strikes or other forms of noncooperation, might seem to employers and to southern communities to have forfeited their right to employment. Under such conditions, employers had a high competitive advantage over northern mills at a time when lack of capital and experience made an advantage especially necessary.[21]

Town leaders looked upon industrialists as pioneers and made numerous concessions to obtain mills. They lowered or rebated taxes, avoided restrictive legislation, and condoned low wages.

The mill village arose partly out of necessity because of the isolation of many of the mills. Mill owners provided housing and other facilities for workers and assumed control over their use. In this way, the mill village offered owners a means of social control over the work force.

> The first generation of mill workers appears to have been, nevertheless, patient and long suffering in spirit. Its members came to mill villages as refugees; for the most part they had been unsuccessful in previous economic

pursuits, and acceptance of industrial employment represented a first stand against pauperism. They had been accustomed to working from sunrise to sunset without murmuring and to expect little in return.[22]

The village church also contributed to labor discipline in mill towns. Liston Pope, in his study of the church in the growth of the textile industry, *Millhands and Preachers,* stated that the contribution of the church lay in the discipline provided through moral supervision of workers: "The greatest contribution of the churches to the industrial revolution in the south undoubtedly lay in the labor discipline they provided through moral supervision of the workers. From the beginning, employers used the church as vehicles of welfare work and supported church programs."[23] Support worked two ways. The church played a role in social control of workers, and the mills supported the church.

Social control led to fear of challenging the system. The size and extent of social control exerted in southern mill towns kept workers, preachers, the media, school boards, doctors, and lawyers quiet. It accounts for lack of organization among workers.

The labor strife that occurred in southern mill towns in the 1920s and 1930s also accounts for the lack of organization in the later part of the twentieth century. Textile workers went on strike in Henderson (1927), Gastonia, Elizabethton, and Marion (1929), Danville (1934), Greensboro (1938), and High Point (1939), and they went on a general strike (1934). Complaints of workers included long hours, low wages, the stretch-out system, paternalistic mill villages, and the denial of the right of workers to organize in order to limit control of employees through collective bargaining. The strikes were unsuccessful. In some places, Gastonia, for example, strikes ended violently. Gastonia officials appealed to the governor of North Carolina for intervention, and the National Guard sent by the governor precipitated violence. During the general strike of 1934, martial law was declared. The National Guard and state troopers helped to break the strike. Sixteen workers lost their lives.

A marked degree of power was exerted over strikers because the mill villages were company controlled.

In a typical mill village in the strike area, the manufacturer owns a good many other things besides places to house his employees. He owns the homes to be sure; he also owns the store, the church, the recreation center, the school, the drugstore, the doctor, the teacher, and the minister. He owns the community and he regulates the life that goes on there after the day's work is over in his mill. He has the power to discharge the worker at the mill, to refuse him credit at his store, to dump a worker's furniture out of a house, to have him expelled from church, to bar his children from school, and to withhold the services of a doctor or a hospital.[24]

These extreme conditions no longer exist, but they help to explain why the southern textile industry is still largely unorganized. Less than 10 percent of the southern workforce is organized. The Amalgamated Clothing and Textile Worker's Union did not begin to fight for recognition of byssinosis as an industrial disease until the early 1970s.

Physicians added to the problem of tardy recognition of byssinosis. The testimony of Dr. I. E. Buff at the 1970 Senate Hearing on Occupational Safety and Health illustrates this problem.

> Then you go and ask the doctors, "How about this byssinosis?" White lung or cotton dust disease, and they don't want to even know about it and they don't even answer you sometimes. Let's talk about the pathology of the disease. There have been very few people who have ever studied the pathology and yet down there you have one after another that die and no one takes the time to take the tissues of the lung to find out what dust is doing to the cells. Because of the control of the big interests, the dominant forces in the four states [North Carolina, South Carolina, Virginia, and Georgia], the doctors are afraid to move.
>
> This is an awful thing. The doctors will not chip the establishment. One doctor told me that if I said somebody had cotton dust disease, they would run me out of town. Another doctor says it would be unhealthy—doctors of medicine.[25]

In general, both physicians and the lay public remained ignorant of the fact that byssinosis was a specific occupational disease. The active denial of the existence of byssinosis by the cotton industry is another reason for nonrecognition, although medical studies for over 100 years clearly established the existence of byssinosis. As late as 1976 at a stockholder's meeting of J. P. Stevens in New York, Board Chairman James D. Finley said, "Byssinosis is something that is alleged to come from cotton dust. It is a word that's been coined, but it has no meaning."[26] In 1969, an editorial stated, "We are particularly intrigued by the term 'Byssinosis,' a thing thought up by venal doctors who attended last year's ILO meetings in Africa where inferior races are bound to be affected by new diseases more superior people defeated years ago."[27] Robert T. P. de Treville, a physician employed by the Industrial Health Foundation, an organization retained by the American Textile Manufacturers Institute to study the problem of byssinosis, said, "At a meeting we held recently of the Foundations' advisors to discuss byssinosis, we concluded that it was best described as a 'syndrome complex' rather than a disease in the usual sense."[28]

Another factor that may have impeded recognition of byssinosis was studies conducted by the United States Public Health Service in the 1930s, which led many to conclude that dust concentrations in cotton textile mills were low and that workers were not at risk. "The low concentrations of dust encountered in this particular study made it apparent that no adverse effect on health was to be anticipated from this source."[29] "In view of the low concentrations of dust encountered, it became apparent that no adverse effect on health was to be anticipated from this source."[30] A textbook published in 1954 states,

> In Britain respiratory disability among cotton textile workers is something of a problem, and the situation seems to be quite different from ours in the United States. Byssinosis, the respiratory complaint resulting from cotton dust inhalation, is almost unknown in the United States, probably because of our relatively high degree of mechanization.[31]

Technological change has also contributed to the problem of byssinosis. Modern technology increased the amount of dust in cotton mills. More than 97 percent of the United States cotton crop is machine harvested.[32] The machines pick up more debris along with the cotton than did manual pickers, and it is crushed and becomes airborne. Speedier machines intensify the risk of byssinosis to workers in the end process by releasing larger quantities of dust. Air conditioning distributes air contaminated in one part of the mill throughout the mill.

All of these circumstances help to explain the tardy recognition of byssinosis as a disease of cotton mill workers in the United States.

RECENT HISTORY

The complex mix of social, political, economic, and technological factors fostered the illusion that byssinosis did not threaten American cotton textile workers. It persisted until the 1960s. In 1960, British investigators McKerrow and Schilling studied two American cotton mills to determine why byssinosis presented a problem in England but not in the United States. McKerrow and Schilling measured airborne dust concentrations of all particle sizes (total dust). In both mills investigated, they found workers with typical histories of byssinosis. The prevalence of symptoms and the fall of $FEV_{1.0}$ in cardroom workers suggested that cardroom dust was active and a potential cause of disability.[33]

Other investigations followed. Dr. Arend Bouhuy's data on cotton workers with respiratory disease in three southern states indicated that disabling byssinosis occurred in the United States and suggested that hundreds of cotton workers suffered from byssinosis in its early as well as late stages. The study states, "The results suggest that it is urgently necessary to investigate the prevalence of byssinosis in the United States." Bouhuys and his colleagues noted that wherever the diagnosis of byssinosis in a small number of cotton workers prompted epidemiological studies, many more workers similarly affected were identified. They concluded that epidemiological studies of the cotton industry were required to establish the true incidence of byssinosis. In 1967 Bouhuys and others conducted an epidemiological study of cotton textile workers in the federal prison in Atlanta: "Sixty-one (29%) of 214 male workers exposed to dust in carding and spinning rooms of a cotton textile mill had byssinosis—that is chest tightness, or cough or both on Mondays during work."[34] Bouhuys suggested that about 17,000 United States cotton textile workers might have byssinosis. Schrag and Gullett conducted an industrial mill study in North Carolina. They found that 63 of 500 cotton textile workers had byssinosis, with prevalence highest in carders. Spinners, weavers, and winders also had the disease.[35]

In 1969, Zuskin's field studies to obtain further data on byssinosis in the United States indicated that symptoms of byssinosis occurred in 25 percent of 59 carders and in 12 percent of 99 spinners employed in two air-conditioned cotton textile mills.

> The prevalence of byssinosis in these mills is higher than would be expected according to previous data, on the basis of the relatively low dust concentrations in the carding and spinning areas. If these prevalence rates are valid for the U.S. cotton textile industry in general, about 8,000 carders and 9,000 spinners have byssinosis (all grades). This estimate included retired workers.[36]

Zuskin also found that total dust concentrations in all work areas exceeded the provisional maximum allowable concentration of 1 mg per cum proposed by Roach and Schilling. It was noted that the composition of dust may be more important than its amount in the air. Technological changes in the cotton industry have caused increasing amounts of trash in baled cotton, owing to the general use of cotton-picking machines. It was speculated that bract, a component of the trash, contains an agent or agents that cause bronchoconstriction and histamine release in human lung tissue.[37] Other recent developments—for example, high-speed carding machines—produce more dust than did older machines operating at lower speeds. "Thus, several technological changes have occurred in the industry which may account for relatively high prevalence of byssinosis in the face of relatively low dust levels."[38] A survey of textile workers in North Carolina indicated that "20% of those workers in preparation areas, 2% of those in yarn processing areas, and 6% of all employees were diagnosed as byssinotic."[39]

The members of the scientific community had generated data indicating that byssinosis did occur in the United States' cotton industry, suggested the need for adequate population studies to assess the full extent of the problem, and stressed the need to prevent byssinosis in the cotton textile industry. The elements of a preventive program were well known: adequate dust removal techniques, periodic check of airborne dust concentrations, preemployment medical examination, and periodic medical surveillance of workers, including use of lung function tests.

The investigations generated interest in byssinosis and led to the National Conference on Cotton Dust and Health held in Charlotte, North Carolina, on May 2, 1970, convened to assess the health problems related to cotton dust exposure. Participants included scientists, representatives of industry, textile unions, government agencies (state and federal), and private physicians. The purpose of the conference was to explore the present knowledge of byssinosis and to summarize the state of the art concerning applicability of health standards and environmental technology. A summary of the conference said,

Byssinosis is a respiratory disease of occupational origin which occurs throughout the world among workers processing cotton, flax, or hemp. Although the symptoms are clearly recognizable, the specific agent responsible for the symptoms has not been defined. The prevalence of the disease can be controlled by reducing the exposure of the workers to the airborne dust associated with the process.[40]

The conference planners designed an educational activity to enlist talent to control and to prevent byssinosis. The conference had five sessions: medical aspects of byssinosis, cotton dust sampling, dust and lint suppression, management's responsibility to people, and foreign and United States investigations.

Recommendations for byssinosis prevention included dust sampling and control, including enclosure, local exhaust and general ventilation, the suggestion that lint-free inhalable dust concentration of a size below 15 micrometers be periodically measured, identification of byssinosis reactors by means of lung function testing, and medical surveillance and management. Priorities included adequate control of dust; cooperation among engineering, medicine, industry and labor to develop solutions; care for partially or totally disabled workers; development of appropriate medical and engineering surveillance; industrywide planning; and application of resources and medical research.[41] During the 1970s, Arend Bouhuys and other concerned scientists continued their research activities.

The 1970 hearings before the Senate Subcommittee on Labor and Public Welfare on proposed legislation for occupational safety and health focused attention upon occupational safety and health problems, including those of the cotton textile industry.

Testimony indicated that no effective government regulation of the hazard of cotton dust existed on either the state or federal level and that workmen's compensation laws were woefully inadequate. George Perkel, spokesman for the Textile Workers Union of America, addressed the need for occupational safety and health legislation to control and prevent diseases such as byssinosis; he cited the studies of Bouhuys and other scientists. State legislators also testified on behalf of needed occupational safety and health legislation.[42]

At the same hearings a statement filed by the American Textile Manufacturers Institute implied that byssinosis was not a problem. For instance, they cited the work of Dr. Philip Enterline and stated that he did not find excessive deaths due to respiratory illness in textile employees: "It would seem that if there were a definite association between exposure to cotton dust and mortality, this association would have been brought out by Dr. Enterline's work." The statement referred to public health service reports in the following manner, "Thus, if as late as 1969 the Public Health Service did not consider cotton or man-made fibers as potentially hazardous there was certainly reason to question statements to the contrary." "We were not surprised," the statement said, "to find out that, in general, industry representatives at these meetings had not heard of byssinosis, nor in most instances had they been aware of any syndrome or symptoms complex such as described in these articles."[43] The American Textile Manufacturers Institute stated in the same document that "it has set out to determine whether or not there is a respiratory problem peculiar to textile workers, and if there is one to ferret out the causes."[44]

In order to make workers, the public, physicians, compensation boards, and industrialists recognize the existence of byssinosis, the Carolina Brown Lung Association organized and held its first meeting in 1975. The Carolina Brown Lung Association, composed mainly of textile workers, sought to organize mass filings of brown lung compensation claims to draw public attention to brown lung (byssinosis) and to identify workers with breathing problems. It organized clinics to educate victims, doctors, the public, and active workers about byssinosis and did much to alert the public about brown lung. In 1977 Carolina Brown Lung Association members testified at the OSHA hearing for a cotton dust standard.

By the mid-seventies, activity generated by the scientific community, public interest groups, and labor created public awareness of byssinosis as an occupationally caused disease. In that time period the new social awareness and a changed attitude toward occupational disease was manifested in the passage of the Occupational Safety and Health Act of 1970. The act created a regulatory agency, OSHA, charged with promulgation and enforcement of workplace safety and health standards.

In 1968, when the Secretary of Labor promulgated under the Walsh-Healy Act a threshold limit value (TLV) for cotton dust of 1,000 μg per cum by adopting the American Conference of Governmental Industrial Hygienists (ACGIH) TLV for cotton dust, the first legally enforceable national exposure limit to cotton dust was established in the United States. The ACGIH, a private standard-setting organization, originally developed the TLV for cotton dust. It was not an enforceable standard. Standards did not become legally enforceable until the Occupational Safety and Health Act of 1970. Then, in accordance with the purpose of the act, to assure as far as possible safe and healthful working conditions for working men and women, Congress vested authority in the Secretary of Labor to set mandatory safety and health standards.

Setting standards involves a number of complexities. One must assume that in order to set a standard there is evidence that presumes predictive validity for the effect that will follow if the standard is exceeded. Along with this scientific component, there is also a value component that asks how much responsibility for health a society will assume. Society must believe that a problem requires attention and must be willing to pay the price to alleviate the problem. Inevitably the following questions are raised: Do the benefits of health protection warrant the expense of an effective standard? Do the benefits of health protection outweigh the risks of lack of protection?

On September 26, 1974, the director of the National Institute for Occupational Safety and Health (NIOSH) submitted a criteria document to the Secretary of Labor. It contained NIOSH recommendations for a new cotton dust standard. On December 27, 1974, OSHA published an "Advanced Notice of Proposed Rule-Making," and requested that interested persons submit their views on cotton dust and the NIOSH Criteria Document. The Textile Workers Union and the North Carolina public interest groups requested that the standard be set at 100 μg/cum rather than at 200 μg/cum, as the Criteria Document had suggested. On December 28, 1976, OSHA published its proposal to revise the existing standard. It called for a permissible exposure limit (PEL) of 200 μg/cum of verticle elutriated cotton dust for all segments of the cotton industry, proposed medical surveillance of workers, detailed the methods of implementation, and contained provisions for employee exposure monitoring, employee training, work practices, record keeping, and methods of exposure control.[45] OSHA held hearings on the proposal in April and May of 1977 in Washington, D.C.; Greenville, Mississippi; and Lubbock, Texas. The record of these hearings contains more than 100,000 pages of documentation and testimony.

The hearings illustrate the complexity of the issues inherent in framing a standard. All parties had their say at the public hearings, often contradicting each other and stating a number of different and opposing viewpoints. Witnesses included manufacturers, employers, lobbyists, lawyers, cotton workers, union representatives, public affairs groups, physicians, scientists, statisticians, economists, and industrial hygienists. The American Textile Manufacturers Association, the Amalgamated Clothing and Textile Workers Union, OSHA, NIOSH, the Carolina Brown Lung Association, the National Organization of Women, and others came together to testify and to represent their constituents' viewpoints. The final standard was based upon the record of the proceedings, which included social, technical, and economic inputs. OSHA's stated viewpoint was that it believed that

The overwhelming scientific evidence in the record supports the finding that cotton dust produces adverse effects among cotton workers. While gaps exist in the understanding of the etiology of respiratory disease caused by cotton dust and their progression from acute to chronic stages, the evidence in the record supports the fundamental connection between cotton dust and various respiratory disorders in both the textile and non-textile industries. Byssinosis is the specific respiratory disease attributable to the action of cotton dust on the respiratory passages.[46]

OSHA believed that the continuing scientific debate over etiology did not detract from the conclusion that cotton dust causes respiratory illness and that enough information existed to warrant controlling the disease byssinosis. The American Textile Manufacturers Institute did not have the same viewpoint, insisting that more research was necessary before a standard could be decided upon and that the standard requested by OSHA was too stringent, not economically feasible, and harmful to the cotton industry.

During the hearings, scientific studies were cited to document the relationship between byssinosis and cotton dust exposure. The American Textile Manufacturers Institute (ATMI) contested work of Merchant that documented dose-response relationships and that provided support for exposure limits[47] and sought to minimize the risk associated with cotton dust exposure.

The ATMI objected to many of the provisions in the proposal, including the definition of cotton dust, permissible exposure limits, exposure monitoring and measurement, methods of compliance, and medical surveillance and monitoring.[48]

The issues of technological feasibility and economic impact were also addressed. The Research Triangle Institute prepared a document on technological feasibility and economic impact, stating as its primary objective "to provide as thorough an assessment as possible of the technological feasibility of controlling cotton dust to exposure levels being considered by OSHA for inclusion in a cotton dust standard, and the economic impact that compliance with the proposed standard may have on industry and the economy."[49] The RTI document stated, "The principal benefit that might be expected to result from implementation of the proposed standard are reduction of the prevalence of byssinosis and cases of byssinosis avoided."[50] With respect to economic impact, RTI said, "Although impact on any one firm cannot be specified in advance, nothing in the RTI study indicated that the cotton textile industry as a whole will be seriously threatened by the impact of the proposed standard for control of cotton dust exposure."[51] At the hearings, Soule, an industrial hygienist, stated, "In summary, for reasons described, it is my opinion that compliance of all segments of the cotton industry with the proposed standard for exposure to cotton dust is technically feasible."[52]

The Council on Wage and Price Stability also prepared a document on the proposed standard and came to a different conclusion: "We believe that the proposed standard could impose excessive costs upon the cotton processing industry unless the standard is carefully formulated to take into account cost-effectiveness considerations and the potential for including positive cost-reducing innovations."[53]

OSHA issued its final cotton dust standard on June 23, 1978. The standard established a permissible exposure limit (PEL) of 200 μg/cum for yarn manufacturing, 750 μg/cum for slashing and weaving operations, and 500 μg/cum for all other processes in the cotton industry and for nontextile industries where there is exposure to cotton dust. The standard also provided for employee exposure monitoring, engineering controls and work practices, respirators, employee training, medical surveillance, signs, and record keeping.

OSHA stressed that the standard was designed to attain the highest degree of health and safety for cotton workers. It did not base exposure levels primarily on cost-benefit analysis.

In making judgments about specific hazards, OSHA is given discretion which is essentially legislative in nature. In setting an exposure limit for a substance like cotton dust, OSHA has concluded that it is inappropriate to substitute cost-benefit criteria for the legislatively determined directive of protecting all exposed employees against material impairment of health or bodily function.[54]

According to OSHA, the standard was justified if compliance costs were not overly burdensome and the benefits appreciable.

On October 24, 1979, the United States Court of Appeals for the District of Columbia issued an opinion upholding the standard. The court also held that the standard was economically and technically feasible. The court rejected the argument that OSHA must conduct a formal cost-benefit analysis before promulgating the standard.

On June 17, 1981, the United States Supreme Court upheld the Cotton Dust Standard. In a five-to-three decision, it rejected arguments by the textile industry that the Cotton Dust Standard was invalid because OSHA failed to show that the cost of compliance was justified by the health benefits to workers. Justice Brennan in the majority statement said,

In effect then, as the Court of Appeals held, Congress itself defined the basic relationship between costs and benefits by placing the "benefit" of worker health above all other considerations save that of making "attainment" of this "benefit" unachievable. Any standard based on a balancing of costs and benefits by the Secretary that strikes a different balance than that struck by Congress would be inconsistent with the command set forth in Section 6(b)(5). Thus, cost-benefit analysis by OSHA is not required by the statute because feasibility is.[55]

Thus, the Supreme Court decision placed priority on worker health in the case of hazard from toxic substances and harmful physical agents. The industry-supported concept of cost-benefit analysis for OSHA relations had been dealt a blow. Perhaps the concept of significant risk would assume more importance than cost-benefit analysis.

In 1981, the same year the Supreme Court upheld the Cotton Dust Standard for the textile industry, the Reagan administration moved to reconsider it. The standard remains fully enforced. In a document prepared for the Office of Technology Assessment book, *Preventing Illness and Injury in the Workplace,* it was concluded that both increasing productivity and compliance with OSHA regulations made important contributions to the modernization of the American textile industry.[56] The following was quoted from a Department of Commerce document:

The United States Occupational Safety and Health Administration dust regulations have had a dramatic effect on . . . processing equipment design and purchasing. Machine suppliers modified equipment to comply with OSHA regulations and this equipment has been accepted on a worldwide basis as well as in the U.S.A. The dust controls have also contributed to much better operating results.[57]

Another quote in the Office of Technology Assessment study stated, "Tougher government regulations on worker's health have, unexpectedly, given the [U.S.] industry a leg up. Tighter dust control rules for cotton plants caused firms to throw out tons of old inefficient machinery and to replace it with the latest available from the world's leading textile machinery firms."[58]

SUMMARY

On the basis of the review of symptoms of respiratory disease and the history of growing awareness and recognition of ailments associated with the manufacture of cotton, it must be concluded that recognition of byssinosis in the United States came very late. In the American South byssinosis existed in an industry and within a social framework that helped to retard the acceptance of byssinosis as an undesirable but controllable disease. The symptoms were recognized as early as 1832; byssinosis was ruled a compensatory disease in England in 1940. As late as the 1960s, United States medical opinion still declared the disease nonexistent in American textile mills. In 1970 the scientific community showed that the problem of respiratory disease peculiar to the cotton industry existed in the United States. Mobilization of public opinion, changes in social attitudes, the new scientific evidence, the Occupational Safety and Health Act, and inputs from groups such as the Carolina Brown Lung Association and the Amalgamated Clothing and Textile Workers Union led to the setting of a new and stringent standard in the cotton textile industry to control and eradicate byssinosis.

Ironically, after decades of society's coming to terms with byssinosis as an occupational disease, implementation of a national standard for control and prevention of the disease was caught up in a debate of benefit-risk consideration. Social factors were previously responsible for inaction. Today, inaction also reflects competing needs in a resource-scarce national economy. Consensus finally exists with respect to the definition, extent, and measures for prevention of byssinosis. Allocation of resources may have to await unequivocal demonstration that the benefits that will accrue from control will exceed the costs of control. Economic factors have become major, identifiable inputs into our social and health policy decisions. Hereafter, byssinosis will be associated with the Supreme Court decision defining the guidelines for risk analysis that will have to precede promulgation and enforcement of a federal occupational health standard.

References

1. Fraser, D. A. and M. C. Battigelli. *Transactions of the National Conference on Cotton Dust and Health.* Chapel Hill: University of North Carolina Press, 1970.
2. Ibid., x.
3. United States Senate, 91st Congress. *Hearings Before a Subcommittee on Labor and Public Welfare.* Washington: U.S. Government Printing Office, 1970.
4. Imbus, H. R. and M. W. Suh. "A Study of 10,133 Textile Workers." *Archives of Environmental Health* 26(1973): 183–191.
5. United States Department of Labor. *An Interim Report to Congress on Occupational Disease.* Washington: U.S. Government Printing Office, 1980.
6. Bouhuys, Arend, et al. "Byssinosis in the United States." *New England Journal of Medicine* 277(1967): 170–175.
7. National Institute for Occupational Safety and Health, *Criteria for a Recommended Standard . . . Occupational Exposure to Cotton Dust.* Washington: U.S. Department of Health, Education, and Welfare, U.S. Public Health Service, Centers for Disease Control, 1974.
8. Thackrah, C. T. *The Effects of Arts, Trades, and Professions and Civic States and Habits of Living, on Health and Longevity.* London: Longman, Rees, Orme, Brown, Green, and Longman, 1832.
9. Ware, N. *The Industrial Worker 1840–1860.* Chicago: Quadrangle Paperbacks, 1964.
10. Josephson, H. *The Golden Threads.* New York: Duell, Sloan and Pearce, 1949. 257.
11. Walkowitz, D. J. *Worker City, Company Town.* Urbana: University of Illinois Press, 1978.
12. Wright, C. D. *Fall River, Lowell and Lawrence: From the Thirteenth Annual Report of the Massachusetts Bureau of Statistics of Labor.* Boston: Rand Avery and Co., 1882.
13. Leach, J. "Surat Cotton as It Bodily Affects Operatives in Cotton Mills." *Lancet* 2(1863): 648–649.
14. Caminita, G. H. et al. *A Review of the Literature Relating to Affections of the Respiratory Tract in Individuals Exposed to Cotton Dust.* Public Health Bulletin 297. Washington: Federal Security Agency, U.S. Public Health Service, 1947.
15. NIOSH, *Criteria,* 23. (Collis quoted in document.)
16. Hill, A. B. *Sickness Amongst Operatives in Lancastershire Cotton Spinning Mills.* London: Medical Research Board, His Majesty's Stationery Office, 1930.

17. Caminita, et al. *Review of the Literature*, 42.
18. Oliver, T. *Diseases of Occupation*. New York: E.P. Dutton and Co., 1908.
19. Caminita, et al. *Review of the Literature*, 26.
20. Ibid., 36.
21. Pope, Liston. *Millhands and Preachers*. New Haven: Yale University Press, 1942.
22. Ibid., 11.
23. Ibid., 29.
24. Tippet, T. *When Southern Labor Stirs*. New York: Jonathan Cape and Harrison Smith, 1931.
25. U.S. Senate. *Hearings*, 1970.
26. Conway, M. *Rise Gonna Rise*. Garden City, N.Y.: Anchor Press, 1979.
27. *American Textile Reporter*. 10 July 1969.
28. Fraser and Battigelli, *Transactions*, 90.
29. Bloomfield, J. J. and W. E. Dreessen, *The Health of Workers in Dusty Trades*. Public Health Bulletin 208 Washington: U.S. Treasury Department, 1933.
30. Britten, R. H. et al. *The Health of Workers in a Textile Plant*, Public Health Bulletin 207 Washington: U.S. Treasury Department, 1933.
31. Drinker, P. and T. Hatch. *Industrial Dust*. Boston: McGraw-Hill, 1954.
32. NIOSH, *Criteria*, 13.
33. McKerrow, C. B. and R. S. F. Schilling. "Pilot Inquiry into Byssinosis in Two Cotton Mills in the United States." *Journal of the American Medical Association* 177(1969): 850–853.
34. Bouhuys, Arend, et al. "Byssinosis in Cotton Textile Workers." *Annals of Internal Medicine* 71(1969): 257–269.
35. Schrag, P. E. and Gullett. "Byssinosis in Cotton Textile Mills." *American Review of Respiratory Diseases* 101(1970): 497–503.
36. Zuskin, E. et al. "Byssinosis in Carding and Spinning Workers." *Archives of Environmental Health* 19(1969): 666.
37. Davies, D. N., ed. *Inhaled Particles and Vapors II*. Oxford, England: Pergamon Press, 1967.
38. Zuskin, et al., "Byssinosis."
39. Merchant, J. A. et al. "Byssinosis and Chronic Bronchitis Among Cotton Textile Workers." *Annals of Internal Medicine* 76(1972): 424–433.
40. Fraser and Battigelli, *Transactions*, iv.
41. "The Status of Byssinosis." *Archives of Environmental Health* 23(1971): 232.
42. U.S. Senate, *Hearings* 587.
43. Ibid., 1005.
44. Ibid., 1006.
45. Federal Register, "Occupational Exposure to Cotton Dust." (Washington: Department of Labor, Occupational Safety and Health Administration, 1978.
46. Ibid., 37352.
47. Ibid., 37355.
48. Proceedings of informal public hearings on proposed standard for exposure to cotton dust (1977): Objections and arguments of American Textile Manufacturers Institute, Inc., concerning proposed standard for exposure to cotton

dust as published by the Department of Labor, Occupational Safety and Health Administration. In *Federal Register* 41(250), 1976.
49. Ibid., 3.
50. Ibid., 8.
51. Ibid., 57.
52. Ibid., 7–8.
53. Ibid., 37.
54. Federal Register, *Occupational Exposure*, 27379.
55. *New York Times,* 18 June 1981.
56. Office of Technology Assessment. *Preventing Illness and Injury in the Workplace*. OTA-H256. Washington: U.S. Congress, Office of Technology Assessment, 1985.
57. Ibid., 87.
58. Ibid., 88.

Index